Advanced Photonic Sciences

Advanced Photonic Sciences

Editor

Dan Hallmen

scitus
academics

Advanced Photonic Sciences
Edited by **Dan Hallmen**

ISBN: 978-1-68117-275-0
Library of Congress Control Number: 2016937447

www.scitusacademics.com
Box No. 4766, 616 Corporate Way,
Suite 2, Valley Cottage,
NY 10989

Printed in United States of America on Acid Free Paper

Preface

Photonics is the science of light emission, transmission, modulation, processing of signals, switching, and amplification and sensing. While it covers all technical light applications across the entire spectrum, most photonic applications are visible and close to infrarouge. The term photonics was developed in the early 1960s by the first practical semiconductor light emitters and in the 1970s by the development of optical fibers. Similar to electronic applications, photon applications use photon. Lighting equipment has several advantages over electricity. Light travels about 10 times the speed of electricity, meaning (among other things) that photonically transmitted data can travel long distances over a fraction of the time. In addition, visible and infrared (IR) beams pass without interacting, unlike electrical currents, so that they do not interfere. A single optical fiber can receive 3-million phone calls simultaneously. Many current or potential photonic applications include: photonic switching, silicon photonics, photonics networks and computer systems. The new emerging photonics field has attracted the attention of many companies, professionals and researchers throughout the world. The importance of this area lies in its application and possible application in almost all scientific and industrial sectors.

Advanced Photonic Sciences offers advanced topics in photonics. The book will be of an assistive tool for experienced professionals, academics and researchers who only begin their professional careers in this field.

Table of Contents

CHAPTER 1

Integrated Microwave Photonic Isolators: Theory, Experimental Realization and Application in a Unidirectional Ring Mode-Locked Laser Diode

Martijn J.R. Heck [1,*], Sudharsanan Srinivasan [2], Michael L. Davenport [2] and John E. Bowers [2]

[1] Department of Engineering, Aarhus University, Aarhus 8200, Denmark

[2] Department of Electrical and Computer Engineering, University of California Santa Barbara, Santa Barbara 93106-9560, CA, USA

ABSTRACT

A novel integrated microwave photonic isolator is presented. It is based on the timed drive of a pair of optical modulators, which transmit a pulsed or oscillating optical signal with low loss, when driven in phase. A signal in the reverse propagation direction will find the modulators out of phase and, hence, will experience high loss. Optical and microwave isolation ratios were simulated to be in the range up to 10 dB and 20 dB, respectively, using parameters representative for the indium phosphide platform. The experimental realization of this device in the hybrid silicon platform showed microwave isolation in the 9 dB–22 dB range. Furthermore, we present a design study on the use of these isolators inside a ring mode-locked laser cavity. Simulations show that unidirectional operation can be achieved, with a 30–50-dB suppression of the counter propagating mode, at limited driving voltages. The potentially low noise and feedback-insensitive operation of such a laser makes it a very promising candidate for use as on-chip microwave or comb generators.

KEYWORDS

hybrid silicon platform; microwave photonics; mode-locked laser diodes; optical isolators; photonic integrated circuits; semiconductor optical amplifiers; silicon photonics

1. INTRODUCTION

The field of microwave photonics (MWP) utilizes photonics to generate, process, transport and detect radio frequencies (RF), typically in the gigahertz range [1]. This approach benefits from the large bandwidth and low loss signal propagation of photonic components, as compared to electronic components and circuits. Examples of successful MWP applications include high-bandwidth wireless links operating at 100 Gbps [2], low noise microwave oscillators [3], clock-free data generation [4] and photonic analog-to-digital converters for high bandwidth operation [5].

Such MWP systems are typically built using fiber-optic and discrete components, limiting their real-world applications, due to lack of robustness, high cost and large size and weight. Integration of MWP systems on a photonic integrated circuit (PIC) can in principle address these concerns and limitations. PIC technology has now evolved to the level that industry and academia have access to mature technology platforms [6]. This led to the field of integrated MWP, where MWP functions are integrated on a PIC [7]. Recent advances, like the ultra-low waveguide losses, realized in the silicon nitride platform, can further unlock this potential, allowing for the almost fiber-like performance of a PIC-based system, e.g., for low noise microwave generation [8].

A main bottleneck for the integration of larger-scale systems on a PIC is the lack of practical integrated optical isolators, e.g., to prevent optical feedback from entering an on-chip laser cavity. For high-end MWP applications, low noise performance is of key importance, but such feedback results in increased phase and relative-intensity noise, increased optical linewidths or even chaotic laser emission [9,10]. Research efforts to realize an integrated isolator are mainly based on the integration of magneto-optic materials onto the PIC platform, to enable a non-reciprocal absorption or phase shift, however with a limited extinction ratio and high insertion loss [11]. Another way to break the symmetry between the forward and backward signals is by applying a time-dependent electro-optic index or transmission modulation, as presented in, e.g., [12,13]. More recently, isolation based on acousto-optic modulation was also shown, though without a clear path towards integration on mature PIC platforms [14].

The time-based asymmetry of a pulsed or MWP signal, propagating through a sequence of optical gates, which open and close periodically, can be used to achieve a difference in loss between forward and backward propagating signals. If a forward propagating oscillating signal passes through an open gate when its amplitude is maximum, its loss is minimal. However, if the spacing and timing between the gates is designed well, a backward propagating signal will not find all gates open and, hence, will incur additional losses. This general concept was theoretically studied in [15], where the self-amplitude modulation of a set of saturable absorption modulators (AMs) was used to create isolation between a pulsed signal and lower power feedback or noise.

In this work, we use a conceptually similar approach, but drive the modulators actively, by applying a time-varying voltage, effectively creating periodic gates. The concept and the simulation of the MWP isolator using realistic parameters, representative for indium phosphide-based active components, will be presented in Section 2. The realization in the hybrid silicon platform will be described in Section 3. Experimentally obtained results will be shown and discussed. In Section 4, we will theoretically investigate a possible promising implementation of the MWP in a ring mode-locked laser (MLL). It will be shown that unidirectional operation of the ring MLL can be achieved. To conclude, in Section 5, we will discuss the wider implementation and application range of these concepts.

2. MWP ISOLATOR CONCEPT AND THEORETICAL STUDY

The concept for an MWP isolator is shown in Figure 1, using two modulators. The modulators are driven with a period T, by varying the voltage over them, and, hence, have a time-dependent transmission. A forward propagating oscillating signal in phase with the transmission modulation of the modulators will experience minimum loss. This figure shows clearly how a backward propagating signal will always encounter at least one modulator in a (partially) blocking state, if the distance between the two modulators corresponds to $0.25 \cdot T \cdot v_g$, with v_g the group velocity of the signal in the system. This means that from the perspective of the backward propagating signal, the two modulator gates are always modulated out of phase with each other, whereas from the perspective of the forward propagating signal, these modulator gates are modulated in phase.

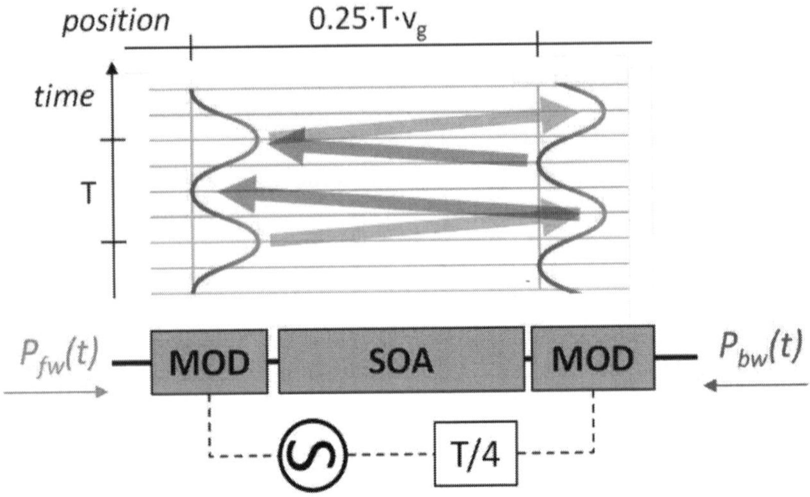

Figure 1. Schematic of the microwave photonics (MWP) isolator concept using modulators (MODs), showing the concatenated MOD-semiconductor optical amplifier (SOA)-MOD array, with the forward propagating signal P_{fw} entering from the left and the backward propagating signal P_{bw} entering from the right. The MOD gates are driven with a $T/4$ time difference, with T the period of the RF signal. The MOD transmission is indicated schematically with the sinusoid, with period T, with red indicating low transmission and green indicating high transmission. The green and red arrows are following the time-position grid and show how a forward propagating signal (green arrow) can pass two open gates, whereas a backward propagating signal (red arrow) always encounters at least on (partially) closed gate. The optical amplifier (SOA) can be used to compensate for overall insertion losses.

Although this concept would work with any type of RF-driven modulator, we limit the discussion to reversely-biased semiconductor optical amplifier (SOA) sections, effectively creating AMs, as was done in [12]. The main reason is the obvious compatibility with the amplifying SOA sections in a PIC, which allows for integration of the MWP isolator after the output of a laser and even into a laser cavity, as will be shown in Section 4. As mentioned above, this is the main rationale for an integrated isolator. From a fabrication perspective, this means that no epitaxial regrowth or overgrowth steps are required, nor techniques like quantum-well intermixing. Experimentally, it was validated that the SOA transmission, more specifically the absorption, changes with applied reverse bias voltage, and hence, it is feasible to use it as an AM [16]. Note that in this implementation, the operating wavelength is well below the bandgap wavelength of the AM, which is different from what is typically called an electro-absorption modulator (EAM), in which case the wavelength is close to the bandgap.

We can model the MWP isolator, as shown in Figure 1, using SOA rate equations for the optical field $A(z,t)$ and the carrier density $N(z,t)$ inside the SOA as presented in, e.g., [12,17]:

$$\frac{\partial A(z,t)}{\partial z} \pm \frac{1}{v_g}\frac{\partial A(z,t)}{\partial t} = \frac{1}{2}\frac{G(z,t)}{1+\varepsilon P(z,t)}A(z,t) - \frac{i}{2}\alpha_N G(z,t)A(z,t) - \frac{1}{2}\alpha_{int}A(z,t) + F_{SE}(z,t) \quad (1)$$

$$\frac{\partial N(z,t)}{\partial t} = \frac{I}{qV} - \frac{N(z,t)}{\tau} - \frac{1}{\Gamma\hbar\omega\sigma}\frac{G(z,t)P(z,t)}{1+\varepsilon P(z,t)} \quad (2)$$

with the gain $G(z,t)$ expressed as a linearized dependency on the carrier density $N(z,t)$:

$$G(z,t) = \Gamma a_N \left(N(z,t) - N_{tr}\right) \quad (3)$$

Here, the optical field $A(z,t)$ is expressed in [$W^{1/2}$], $P(z,t)$ is the optical power in [W], ε is the nonlinear gain compression (0.2 W^{-1} for SOA, 0 for AM, *i.e.*, reverse bias), α_N is the linewidth enhancement factor (3 for SOA, 1 for AM), α_{int} = 20 $cm^{-1}$the passive loss, $F_{SE}(z,t)$ a Langevin noise term representing spontaneous emission, I the injection current, q the elementary charge, V the volume of the active area, τ the carrier recovery time (300 ps for SOA, 5 ps for AM), Γ = 0.3 the confinement factor, $\hbar$ is Planck's constant, ω the optical angular frequency, σ the mode cross section, a_N the differential gain coefficient (3×10^{-20} m^2 for SOA, 12×10^{-20} m^2 for AM) and N_{tr} the transparency carrier density (0.3×10^{18} cm^{-3} for SOA, 0.7×10^{18} $cm^{-3}$for AM). These parameters are representative for the indium phosphide technology and have been experimentally verified, e.g., [13]. The absorption dependency of the AM on the bias voltage is modeled using a voltage-dependent transparency carrier density $N_{tr} = N_{tr}(V(t))$, as proposed in [18] and experimentally verified in [13]. These equations are numerically solved using a spatial resolution of Δz = 5 μm, corresponding to a time resolution of Δt = 60 fs.

Using this model, the feasibility of the MWP isolator can be explored. The transmission of the forward and backward propagating 20-GHz MWP fields for an isolator with 100 μm-long AMs is shown in Figure 2a. Obviously, the AM driving voltage needs to be in phase with the MWP signal oscillator for maximum transmission, as shown by the maximum around three radians drive offset. Note that this is a matter of calibrating the model, or any real-world system, and represents a relative value only. Since a system integrated on a PIC has known path lengths, the required drive offset is known beforehand. On the other hand, the counter-propagating signal has in practice an unknown phase with respect to the AMs, as reflections can originate from anywhere downstream of the circuit. However, as can be seen in Figure 2a, this transmission is flat and below the peak transmission of the forward signal, indicating the isolation capability. Note that this flat transmission also means that isolation from continuous-wave counter-propagating signals is achieved.

In Figure 2b, the isolation is plotted for various AM lengths and various driving voltage amplitudes of 3–9 V (corresponding to $N_{tr} = (0.1–0.3) \times 10^{18}$ cm^{-3}; [13]). As can be expected, isolation increases with AM length and driving voltage amplitude. However, it has to be noted that such driving voltages are quite high, requiring between 100 mW and 1 W of RF power on the AM, assuming a 50-Ω termination. Based on these results, it can be seen that modest optical isolation ratios up to 10 dB can be achieved, which corresponds to close to a 20-dB microwave isolation, as most of the power is in the 20-GHz microwave fundamental harmonic, with up to 20% in the 40-GHz harmonic, due to the nonlinear transmission.

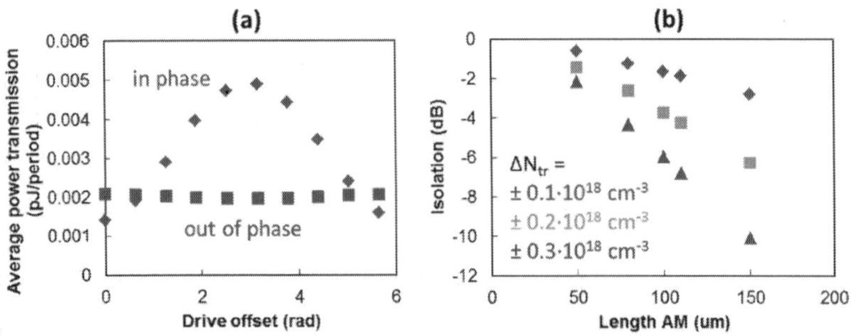

Figure 2. (**a**) Power transmission of a 1-mW peak power 20-GHz MWP signal through an isolator with a total length of 1.2 mm and with two absorption modulator (AM) sections of 100 μm. The bias amplitude corresponds to $\pm 0.2 \times 10^{18}$ cm^{-3}. Referring toFigure 1, the forward transmission (red) and reverse, backward transmission (blue) are shown, as a function of the phase difference between the MWP signal and the AM driving voltage. (**b**) Maximum difference between forward and backward propagating signals for various AM lengths and driving voltages (indicated with N_{tr}).

3. EXPERIMENTAL REALIZATION IN HYBRID SILICON TECHNOLOGY

The MWP isolator was fabricated in the hybrid silicon platform [19]. In this platform, an indium phosphide layer is bonded to a processed silicon photonic chip and then processed further, ensuring lithographic alignment accuracy. Silicon photonic components, such as waveguides, splitters and multiplexers, can be combined in this platform with hybrid waveguides, where the optical mode overlaps with the indium phosphide layers. In this way, active optical functionalities, such as gain, can be achieved. A schematic of the device cross-section is shown in Figure 3a, where it can be seen that the optical mode propagates through an AM-SOA-AM sequence. The tapered mode converters from the silicon waveguide to the hybrid waveguide are electrically pumped, hence the small SOA sections at the start and end of the device [20]. For details about the fabrication process, we refer to [16]. The chip is diced, polished and anti-reflection coated for characterization.

The experimental setup to characterize the MWP isolator is shown in Figure 3c. An MWP signal is generated by an oscillator that is synchronized with the AM drivers. Two microwave phase shifters allow for independent tuning of the phase of the driving voltage on the AMs, compared to the MWP signal. The optical carrier wavelength is 1583 nm, as the AM has maximum dynamic range here, expressed as absorption change as a function of applied voltage. This was determined by looking at the amplified spontaneous emission spectrum of the SOA through the AM, on which the bias was varied, as shown in Figure 4a. The MWP isolator insertion loss is 4.4 ± 0.5 dB, which of course depends on the SOA bias, in this case 200 mA. The overall active device length is 1.4 mm, with two 100-μm long AMs, positioned with a 1.0-mm spacing.

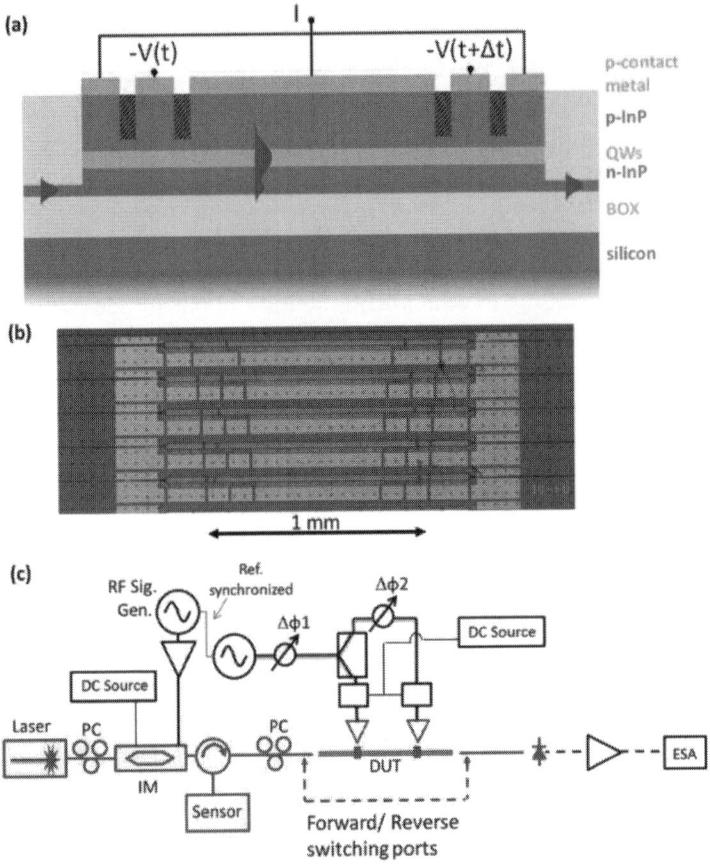

Figure 3. (**a**) Schematic of the hybrid silicon MWP isolator, showing the optical field (red) as it propagates through the silicon waveguide, hybrid silicon III/V waveguide and the silicon waveguide again. Hydrogen-implantation of the p-doped InP cladding provides electrical isolation between the forward-biased SOA and the reverse-biased AM. N-contacts are to the side and not shown in this cross-section. (**b**) Microscope picture of an array of realized devices. (**c**) Experimental setup, showing the MWP signal generation using an intensity modulator (IM, $V_\pi = 3.6$ V), driven by a signal generator (RF Sig. Gen.). This signal generator is synchronized with another signal generator that drives the two AMs. Two microwave phase shifters ($\Delta\varphi_{1,2}$) are used to control the phase. The signal is recorded using an electrical spectrum analyzer (ESA) connected to a 40-GHz bandwidth photodiode and a 20-dB RF amplifier. Polarization controllers (PCs) are used to ensure transverse-electric polarization of the field inside the device under test (DUT). Lensed fibers are used to couple the light in and out of the device. Electrical connections are shown in black, optical connections in red. An estimated power of 12.5 dBm is used to drive each of the AMs. The SOA is biased at 200 mA. The circulator with the sensor is used for fiber to chip alignment.

Figure 4b shows the 20-GHz MWP signal transmission with $\Delta\varphi_1$ set to drive one AM in phase with the MWP signal, while tuning the phase of the voltage on the other AM by $\Delta\varphi_2$ (Figure 3b). As can be seen, the received RF power varies with changing the phase on the second AM. Using this plot as a calibration, it can be seen that at $\Delta\varphi_2 = 50° \pm 10°$, the two AMs are driven at $\Delta T = T/4$, as explained in Figure 1, with $T = 50$ ps. This means maximum transmission for the forward signal and minimum transmission for the counter-propagating signal. At $\Delta\varphi_2 = 230° \pm 10°$, this situation is reversed, as expected, and represents $\Delta T = -T/4$. From this plot, it can be seen that 22-dB and 9-dB microwave isolations have been achieved, which agrees well with the optical isolation ratios up to 10 dB, obtained using our simulation. This comparison should be considered qualitative, as a complete set of SOA and AM parameters for the hybrid silicon platform is not available, and the simulation used parameters from the monolithic indium phosphide technology. However, since the active part of the hybrid waveguide is indium phosphide and since the simulation tool uses parameter values experimentally obtained with an indium phosphide SOA, it can be assumed that these outcomes should be close.

We note that ideally, the operation of the device should be symmetric for forward and reverse propagation. From Figure 4b, it can be seen that this is not the case in the experiments. We hypothesize that due to processing non-uniformities, most notably variations in contact resistance over the length of the device, the AMs do not operate identically. In the linear regime, that would still not matter, but as the AM and SOA are nonlinear devices, *i.e.*, their transmission is power dependent, a difference between the two AMs would lead to a difference between the forward and reverse signals, even when both AMs are driven in phase.

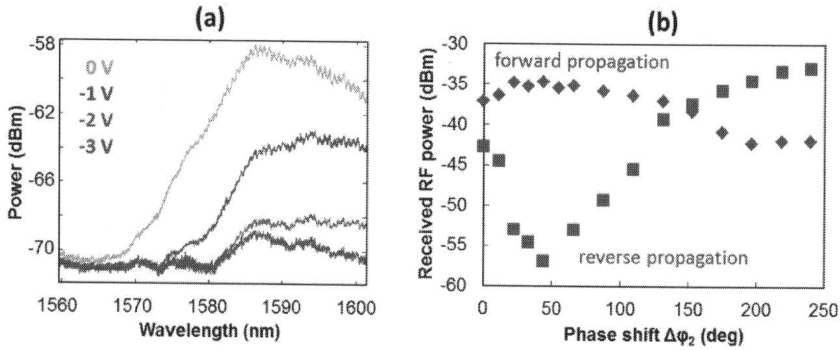

Figure 4. (**a**) Amplified spontaneous emission spectra, obtained by biasing the SOA with a constant current and measured after propagation through an AM. Bias voltage on the AM is varied from 0–3 V. (**b**) The RF power measured with the electrical spectrum analyzer for the forward and backward propagating MWP signals as a function of the phase difference between the two AM drive voltages.

4. APPLICATION OF THE MWP ISOLATOR IN UNIDIRECTIONAL RING MODE-LOCKED LASERS

The theoretical and experimental results show clearly the potential for isolation, though with a somewhat limited extinction ratio up to ~10 dB optical and ~20 dB electrical. One possible application could be to use this MWP isolator inside a ring laser cavity. It is well known that a small periodic variation in the gain can cause mode-locking, *i.e.*, the laser operates in a pulsed mode [21]. Such MLLs have been realized in various configurations, but ring lasers are particularly attractive due to the lack of mirrors and, hence, the ease of integration on a PIC. However, such ring MLLs are intrinsically bi-directional [18], which leads to lower output efficiency and higher amplitude and phase noise due to cross-gain modulation. It has been shown theoretically and experimentally that by careful MLL cavity design, the power ratio in two counter propagating fields can be controlled, however only to a limited extent [22,23].

In this section, we explore theoretically the feasibility of using the MWP isolator to force the MLL into a unidirectional regime. We use the model presented above (Equations (1)–(3)), extended with a 5-THz SOA gain bandwidth filter, to limit the pulse shortening. The model is implemented using a similar approach as in [19]. Figure 5a shows the layout of a ring MLL with a dual AM-phased drive, which means that the MLL is effectively hybridly mode-locked, as the AMs are saturable absorbers and this saturation contributes to the pulse shaping mechanism. The length of the ring determines the period T of the output pulses, and the AMs are placed with a separation corresponding to $T/4$. Figure 5b,c shows the simulated clockwise (CW) and counter-clockwise (CCW) optical output fields for a 20-GHz ring MLL and with an AM drive delay of 12.5 ps and −12.5 ps, respectively. As can be seen, the ring MLL operates unidirectionally with the correct phase of the drive voltage on the AMs. The isolation or suppression of the counter propagating field depends on the MLL operating conditions. In Figure 6, this is shown for a MLL with varying drive voltage amplitude, here again expressed by varying the amplitude of the oscillation of the transparency carrier density N_{tr}. The suppression of the counter propagating field ranges from 30 dB–50 dB, for increasing drive voltage amplitude.

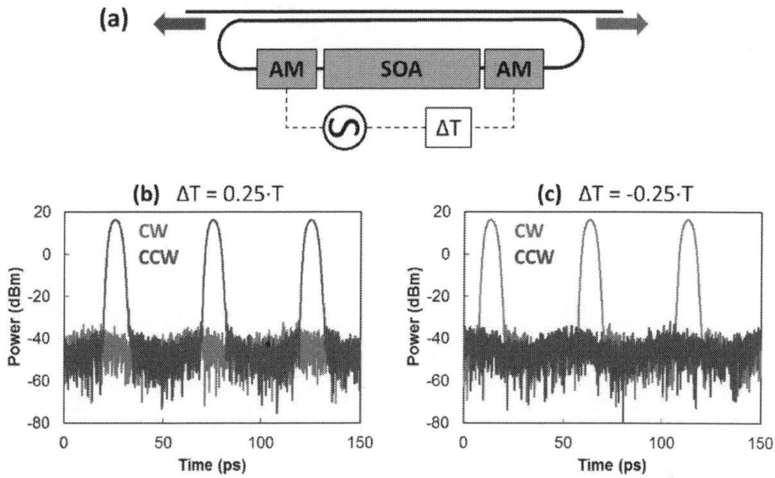

Figure 5. (**a**) Schematic layout of the unidirectional mode-locked laser (MLL), showing an oscillator driving the two AMs, with a relative delay of ΔT. A directional coupler couples light out of the ring into the clockwise (CW, blue) or counterclockwise (CCW, red) output waveguide. The simulated output of a 20-GHz MLL is shown for (**b**) $\Delta T = 12.5$ ps and (**c**) $\Delta T = -12.5$ ps. The SOA length is 800 μm, and the AMs have a length of 25 μm. Power output coupling is 0.5 per roundtrip. The simulation parameters are $\Delta N_{tr} = 0.7 \times 10^{18}$ cm^{-3} and $I_{SOA} = 80$ mA.

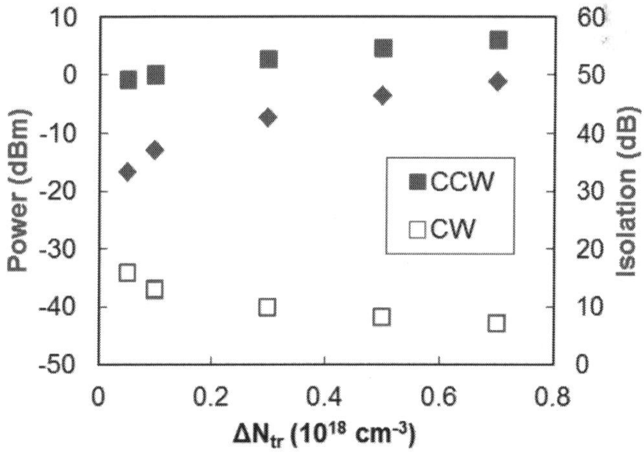

Figure 6. Average output power for the clockwise (CW) and counter-clockwise (CCW) signals, assuming $\Delta T = 12.5$ ps in Figure 5a, as a function of the modulation amplitude ΔN_{tr} of N_{tr}. The isolation ratio is also shown (blue). The SOA injection current is 80 mA. Further MLL design parameters are as in Figure 5.

In conclusion, we have shown the feasibility of an integrated unidirectional ring MLL, using our experimentally-verified concept of an MWP isolator. Apart from a more efficient output, the lack of cross-gain modulation and spatial hole burning will likely lead to lower phase and amplitude noise. Furthermore, it can be hypothesized that the laser is less sensitive to optical feedback, as this feedback is injected in the suppressed counter propagating mode. Major applications can be foreseen in on-chip low noise microwave generation and in on-chip comb generation for wavelength-division multiplexed systems.

5. CONCLUSION AND OUTLOOK

In this paper, we have presented the concept of a novel integrated MWP isolator and its first experimental realization in the hybrid silicon platform. Optical and microwave isolation ratios were simulated to be in the range up to 10 dB and 20 dB, respectively, using parameters representative for the indium phosphide platform. The experimental realization of this device in the hybrid silicon platform showed microwave isolation in the 9 dB–22 dB range.

Such isolation ratios are barely sufficient for on-chip use, for example to shield a laser from optical feedback. As was argued in [24], optical isolation of at least 20 dB is required. Options to enhance the MWP isolator include the concatenation of more AM gates and using more square-shaped driving pulses, which would also reduce higher harmonic generation.

In an alternative implementation, when using this MWP isolator inside a resonant cavity, the isolating effect is strongly enhanced. We have presented a design study on the use of MWP isolators inside a ring MLL cavity. Simulations show that unidirectional operation can be achieved, with a 30–50-dB suppression of the counter propagating mode, at limited driving voltages. The potentially low noise and feedback-insensitive operation of such a laser makes it a very promising candidate for use as on-chip microwave or comb generators.

The simulations were based on the indium phosphide material system, and an implementation using reversely-biased SOA sections or AMs was chosen. This choice offers a practical path to integration with (forward biased) SOAs, for example to realize the unidirectional ring MLLs that were presented in this work. Alternative approaches for the AMs can include Mach–Zehnder interferometer modulators in the silicon photonics platform, electro-absorption modulators in the indium phosphide and hybrid silicon platforms or even forward-biased SOAs as gates. The choice will depend on all trade-offs that are required for PIC design and fabrication. Choosing modulators with a higher extinction ratio will improve the isolation ratio. Another important aspect is the optical bandwidth that is required. Mach–Zehnder interferometer modulators and forward-based SOAs are broadband, having bandwidths of at least tens of nanometers. Electro-absorption modulators and reversely-biased SOAs have a very wavelength-dependent transmission and are hence narrow band, typically only a few nanometer, e.g., as shown in [25] and in Figure 4.

The experimental realization on the hybrid silicon platform is a relevant step, as this platform has been shown to have interesting applications in the field of MWP [26]. Although one can argue that similar PIC functionalities can be achieved using a monolithic indium phosphide platform, the hybrid silicon platform offers the potential of economies of scale, in terms of fabrication on 6" or even 8" substrate wafers and the compatibility with the mature CMOS manufacturing infrastructure. This also includes the required integration with electronic driving and control circuitry, which can be done in a package [27] or fully monolithically on the same chip [28]. Such an approach is necessary for large volume applications that can be envisioned for MWP, which include, for example, 5G wireless communications infrastructure.

ACKNOWLEDGMENTS

The authors would like to thank Greg Fish and Erik Norberg from Aurrion Inc. for helpful discussions. This work was supported by the DARPA EPHI Project. M.J.R.H. acknowledges partial funding from Det Frie Forskningsråd through the mmW-SPRAWL project (DFF-4005-00246).

AUTHOR CONTRIBUTIONS

M.J.R.H. conceived of the idea and did the simulations and mask design. M.L.D. developed the fabrication process and fabricated the devices. S.S. performed the experimental characterization. M.J.R.H. and J.E.B. supervised the project. M.J.R.H. wrote the paper, with input from all other co-authors.

CONFLICT OF INTEREST

The authors declare no conflict of interest.

REFERENCES

1 Yao, J. Microwave photonics. *J. Lightw. Technol.* **2009**, *3*, 314–335.

2 Pang, X.; Caballero, A.; Dogadaev, A.; Arlunno, V.; Borkowski, R.; Pedersen, J.S.; Deng, L.; Karinou, F.; Roubeau, F.; Zibar, D.; *et al.* 100 Gbit/s hybrid optical fiber-wireless link in the W-band (75–110 GHz). *Opt. Express* **2011**, *19*, 24944–24949.

3 Yao, X.S.; Maleki, L. Optoelectronic microwave oscillator. *JOSA B* **1996**, *13*, 1725–1735.

4 Chi, Y.-C.; Lin, G.-R. A self-started laser diode pulsation based synthesizer-free optical return-to-zero on-off-keying data generator. *IEEE Trans. Microwave Theory Tech.* **2010**, *58*, 2292–2298.

5 Ghelfi, P.; Laghezza, F.; Scotti, F.; Serafino, G.; Capria, A.; Pinna, S.; Onori, D.; Porzi, C.; Scaffardi, M.; Malacarne, A.;*et al.* A fully photonics-based coherent radar system. *Nature* **2014**, *507*, 341–345.

6 Smit, M.K.; Leijtens, X.; Bente, E.; Van der Tol, J.; Ambrosius, H.; Robbins, D.; Wale, M.J.; Grote, N.; Schell, M. A generic foundry model for InP-based photonic ICs. In Proceedings Optical Fiber Communication Conference OM3E-3, Los Angeles, LA, USA, 4–8 March 2012.

7 Marpaung, D.; Roeloffzen, C.; Heideman, R.; Leinse, A.; Sales, S.; Capmany, J. Integrated microwave photonics. *Laser Photonics Rev.* **2013**, *7*, 506–538.

8 Heck, M.J.R.; Bauters, J.F.; Davenport, M.L.; Spencer, D.T.; Bowers, J.E. Ultra-low loss waveguide platform and its integration with silicon photonics. *Laser Photonics Rev.* **2014**, *8*, 667–686.

9 Tkach, R.W.; Chraplyvy, A.R. Regimes of feedback effects in 1.5 μm distributed feedback lasers. *J. Lightw. Technol.* **1986**, *4*, 1655–1661.

10 Barbarin, Y.; Bente, E.A.J.M.; Heck, M.J.R.; Den Besten, J.H.; Guidi, G.; Oei, Y.S.; Binsma, J.J.M.; Smit, M.K. Realization and modeling of a 27-GHz integrated passively mode-locked ring laser. *IEEE Photonics Technol. Lett.* **2005**, *17*, 2277–2279.

11 Mizumoto, T.; Takei, R.; Shoji, Y. Waveguide optical isolators for integrated optics. *IEEE J. Quant. Elect.* **2012**, *48*, 252–260.

12 Galland, C.; Ding, R.; Harris, N.C.; Baehr-Jones, T.; Hochberg, M. Broadband on-chip optical non-reciprocity using phase modulators. *Optics Express* **2013**, *21*, 14500–14511

13 Tzuang, L.D.; Fang, K.; Nussenzveig, P.; Fan, S.; Lipson, M. Non-reciprocal phase shift induced by an effective magnetic flux for light. *Nat. Photonics* **2014**, *8*, 701–705.

14 Li, E.; Eggleton, B.J.; Fang, K.; Fan, S. Photonic Aharonov-Bohm effect in photon-phonon interactions. *Nat. Commun.* **2014**, *5*, 1–5.

15 Heck, M.J.R.; Bente, E.A.J.M.; Barbarin, Y.; Lenstra, D.; Smit, M.K. Monolithic Semiconductor Waveguide Device Concept for Picosecond Pulse Amplification, Isolation and Spectral Shaping. *IEEE J. Quant. Elect.* **2007**, *43*, 910–922.

16 Heck, M.J.R.; Bente, E.A.J.M.; Barbarin, Y.; Fryda, A.; Jung, H.-D.; Oei, Y.S.; Nötzel, R.; Lenstra, D.; Smit, M.K. Characterization of a monolithic concatenated SOA/SA waveguide device for picosecond pulse amplification and shaping. *IEEE J. Quant. Elect.* **2008**, *44*, 360–369

17 Jones, D.J.; Zhang, L.M.; Carroll, J.E.; Marcenac, D.D. Dynamics of monolithic passively mode-locked semiconductor lasers. *IEEE J. Quant. Elect.* **1996**, *31*, 1051–1058.

18 Bandelow, U.; Radziunas, M.; Vladimirov, A.; Hüttl, B.; Kaiser, R. 40 GHz mode-locked semiconductor lasers: Theory, simulations and experiment. *Opt. Quant. Elect.* **2006**, *38*, 495–512.

19 Heck, M.J.R.; Bauters, J.F.; Davenport, M.L.; Doylend, J.K.; Jain, S.; Kurczveil, G.; Srinivasan, S.; Tang, Y.; Bowers, J.E. Hybrid silicon photonic integrated circuit technology. *IEEE J. Sel. Top. Quant. Elect.* **2013**, *19*.

20 Kurczveil, G.; Pintus, P.; Heck, M.J.R.; Peters, J.D.; Bowers, J.E. Characterization of insertion loss and back reflection in passive hybrid silicon tapers. *IEEE Photonics J.* **2013**, *5*.

21 Barbarin, Y.; Bente, E.A.J.M.; Heck, M.J.R.; Oei, Y.S.; Nötzel, R.; Smit, M.K. Characterization of a 15 GHz integrated bulk InGaAsP passively modelocked ring laser at 1.53 μm. *Opt. Express* **2006**, *14*, 9716–9727.

22 Bente, E.A.J.M.; Barbarin, Y.; Heck, M.J.R.; Smit, M.K. Modeling of integrated extended cavity InP/InGaAsP semiconductor modelocked ring lasers. *Opt. Quant. Elect.* **2008**, *40*, 131–148.

23 Tahvili, M.S.; Barbarin, Y.; Leijtens, X.J.M.; De Vries, T.; Smalbrugge, E.; Bolk, J.; Ambrosius, H.P.M.M.; Smit, M.K.; Bente, E.A.J.M. Directional control of optical power in integrated InP/InGaAsP extended cavity mode-locked ring lasers. *Opt. Lett.* **2011**, *36*, 2462–2464.

24 Heck, M.J.R. Grating Coupler Enabled Optical Isolators and Circulators for Photonic Integrated Circuits. *IEEE J. Sel. Top. Quant. Elect.* **2015**, *21*, 1–9.

25 Tang, Y.; Chen, H.-W.; Jain, S.; Peters, J.D.; Westergren, U.; Bowers, J.E. 50 Gb/s hybrid silicon traveling-wave electroabsorption modulator. *Opt. Express* **2011**, *19*, 5811–5816.

26 Fish, G. Heterogeneous photonic integration for microwave photonic applications. In Proceedings Optical Fiber Communication Conference and Exposition and the National Fiber Optic Engineers Conference (OFC/NFOEC), Anaheim, CA, USA, 17–21 March 2013.

27 Dupuis, N.; Lee, B.G.; Proesel, J.E.; Rylyakov, A.; Rimolo-Donadio, R.; Baks, C.W.; Ardey, A.; Schow, C.L.; Ramaswamy, A.; Roth, J.E.; *et al.* 30-Gb/s Optical Link Combining Heterogeneously Integrated III-V/Si Photonics With 32-nm CMOS Circuits. *J. Lightw. Technol.* **2015**, *33*, 657–662.

28 Thomson, D.J.; Porte, H.; Goll, B.; Knoll, D.; Lischke, S.; Gardes, F.Y.; Hu, Y.; Reed, G.T.; Zimmermann, H.; Zimmermann, L. Silicon carrier depletion modulator with 10 Gbit/s driver realized in high-performance photonic BiCMOS. *Laser Photonics Rev.* **2014**, *8*, 180–187

CHAPTER 2

Hybrid Photonic Integration on a Polymer Platform

Ziyang Zhang [1,*], David Felipe [1], Vasilis Katopodis [2], Panos Groumas [2], Christos Kouloumentas [2], Hercules Avramopoulos [2], Jean-Yves Dupuy [3], Agnieszka Konczykowska [3], Alberto Dede [4], Antonio Beretta [4], Antonello Vannucci [4], Giulio Cangini [5], Raluca Dinu [5], Detlef Schmidt [1], Martin Moehrle [1], Patrick Runge [1], Jung-Han Choi [1], Heinz-Gunter Bach [1], Norbert Grote [1], Norbert Keil [1] and Martin Schell [1]

[1] Fraunhofer Heinrich Hertz Institute (HHI), Einsteinufer 37, Berlin 10587, Germany

[2] Photonic Communications Research Laboratory, National Technical University of Athens, Zografou, Athens 15573, Greece

[3] III-V Lab, Joint lab between Alcatel-Lucent Bell Labs, Thales-TRT and CEA/Leti, Route de Nozay, Marcoussis 91460, France

[4] Linkra Srl, Via S. Martino 7, Agrate Brianza (MB) 20864, Italy

[5] GigOptix Inc., 19910 North Creek Parkway Suite 100, Bothell, WA 98011, USA

ABSTRACT

To fulfill the functionality demands from the fast developing optical networks, a hybrid integration approach allows for combining the advantages of various material platforms. We have established a polymer-based hybrid integration platform (polyboard), which provides flexible optical input/ouptut interfaces (I/Os) that allow robust coupling of indium phosphide (InP)-based active components, passive insertion of thin-film-based optical elements, and on-chip attachment of optical fibers. This work reviews the recent progress of our polyboard platform. On the fundamental level, multi-core waveguides and polymer/silicon nitride heterogeneous waveguides have been fabricated, broadening device design possibilities and enabling 3D photonic integration. Furthermore, 40-channel optical line terminals and compact, bi-directional optical network units have been developed as highly functional, low-cost devices for the wavelength division multiplexed passive optical network. On a larger scale, thermo-optic elements, thin-film elements and an InP gain chip have been integrated on the polyboard to realize a colorless, dual-polarization optical 90° hybrid as the frontend of a coherent receiver. For high-end applications, a wavelength tunable 100Gbaud transmitter module has been demonstrated, manifesting the joint contribution from the polyboard technology, high speed polymer electro-optic modulator, InP driver electronics and ceramic electronic interconnects.

KEYWORDS

hybrid photonic integration; photonic integrated circuits; polymer waveguides

1. INTRODUCTION

Photonic integration endeavors to bring multiple optical functions into a compact device with a small footprint, low power consumption and reliable performance, analogous to electronic integration. However, contrary to the situation in the electronic industry where silicon CMOS dominates, photonic devices nowadays do not tend to converge to one major material system. Light sources are mostly realized using III-V semiconductor-related materials. Modulators have been dominated by lithium niobate ($LiNbO_3$), but have been gradually taken over by InP and silicon-on-insulator (SOI) platforms. Detectors are fabricated using III-V semiconductors and Si-Ge materials. Waveguide networks, filters, and other passive optical functions can be realized in low-cost platforms such as polymer, silica and silicon nitride planar lightwave circuits (PLCs). Hybrid photonic integration allows for selecting individual components from the respective best-suited material platform and for assembling them in a common motherboard. These components can be tested prior to the assembly. In this way, high freedom of device performance and yield optimization can be realized, leading to a relatively high cost-efficiency.

With the help of the automatic alignment tools and machine-assisted assembly equipment, hybrid integration can offer the benefits of micro-optical packaging at reduced costs, potentially higher overall module performance and higher reliability. Its versatility makes it particularly useful for low to medium production volumes.

In the last decade, polymer-based photonic devices and the hybrid integration platform have witnessed significant advancement [1,2,3]. Multi-core polymer waveguides and multi-step 45° mirror arrays have been fabricated as a bridging devices toward spatial-division-multiplexed (SDM) transmission systems [4,5,6]. Silicon nitride (SiN_x) has been introduced as high-index core material and related thermo-optic devices have been reported [7,8,9,10]. Polymer arrayed waveguide gratings (AWGs) coupled with 40-channel laser diodes (LDs) and photo detectors (PDs) have been implemented as optical line terminals (OLTs) in the wavelength-division-multiplexed passive optical network (WDM-PON) [11,12]. Clever arrangement of thermally tunable Bragg gratings, vertically inserted thin-film element (TFE), on-chip U-groove and 45° mirror has led to the assembly of a compact, bi-directional optical network unit (ONU) [13]. Ultrafast electro-optic polymer modulators, integrated with InP-based light sources and driver electronics, have been packaged into transmitter modules, providing serial 100 Gb/s connectivity in data centers and metro-area optical networks [14,15,16,17,18,19].

In this work, a review is performed summarizing the latest development of the polyboard technology. From waveguides to optical assemblies and finally to packaged modules, details of the heterogeneous/ hybrid integration technology are revealed, uncovering the integration flexibility, scalability, and cost-effectiveness this platform offers. Moreover, functionalities such as wavelength tunability and polarization control can be realized in a simple and clean manner, making this platform highly attractive in applications where colorless and polarization-multiplexed operations are required.

2. POLYMER WAVEGUIDES

Both ridge waveguides and channel waveguides can be fabricated on polymer platforms. The square-shaped core of a channel waveguide can be adopted to minimize the waveguide birefringence. The refractive index can be chosen from ~1.4 to ~1.7, depending on the chemical composition. The dimension of the core measures a few micrometers for single mode operation. Optical polymer layers can be formed by spin-coating and subsequent UV and temperature curing. Silicon wafers are usually taken as substrate, though the technology is also applicable on glass plates, ceramic and InP wafers. The structuring of the waveguide follows mainly three methods [20]: (1) standard semiconductor processing methods with photolithography and reactive ion etching; (2) direct UV lithographic patterning followed by wet-chemical rinsing; and (3) soft molding and imprinting. In this work, the first fabrication method has been used. The technology details have been published in [2,5,7]. Metals can be inserted virtually anywhere in the vicinity of the waveguide to function as heater electrodes for thermally tunable devices but also as RF electrodes to drive the electro-optic (EO) polymer-based modulator.

The fabrication process can be repeated to stack up the polymer waveguide layers into a multi-core structure [5], as shown in Figure 1a,b. The waveguide features a square-core of 3 μm × 3 μm and a core/cladding index contrast of 1.48/1.45. A bending radius of 1.5 mm of such a waveguide will lead to a loss ~0.12 dB for a 90° bend. The five-layer polymer waveguides shown in Figure 1a add up to a total thickness of 108 μm, within which the bottom cladding is 20 μm directly above the silicon substrate, the inter-waveguide-layer distance is 17.32 μm (4 times), and the upper cladding is 15.7 μm below air. The waveguide loss among all five layers is below 1 dB/cm and the crosstalk between the waveguides at the 20-μm hexagonal lattice is below −40 dB after a propagation distance of 3.8 cm. This technology can be exploited to fabricate three-dimensional photonic integrated circuits (PICs) with high vertical density.

On a polymer platform, 45° facets can be created by applying a fast-rotating angled dicing blade and sawing through the polymer chip [2]. The surface roughness can be minimized by the intrinsic "polishing" during sawing. The high concentration of small diamond grains on the blade tip, fast rotation speed and slow feeding speed can help generate a smooth surface. To improve the reflectivity, the 45° surface should be coated with a thin layer of metal, typically Ti and Au, by directed metal evaporation. Apart from the design and processing simplicity, such 45° deflection mirrors offer much broader coupling bandwidth, less polarization sensitivity, and less beam divergence, when compared to the vertical grating couplers. The 45° groove is usually filled with index matching epoxy so that the back reflection on the straight facet can be suppressed and the beam divergence in the free space can be reduced.

The 45° mirrors can be arrayed by varying the dicing depths to tap out light from each of the multi-layer waveguide cores separately, as indicated in Figure 2. The depth of the mirrors must be calculated and well controlled so that the tip reaches, in the ideal case, the middle point between the two core layers. Light from each layer can then be subsequently extracted by the mirror array. To ensure that the metal coating reaches the very bottom of the deep mirror, while staying clear from the straight waveguide facet, thermal evaporation with well-defined deposition directivity must be used. The tilt angle must be carefully calculated. The waveguides can fan out with a larger in-plane pitch and the lateral distance between the mirrors can be adjusted to allow the attachment of either separate 1D planar PD bars or a common 2D PD array. The attachment can be done passively and automatically by a pick-and-placer. No extra submount is needed for the backside illuminated PDs.

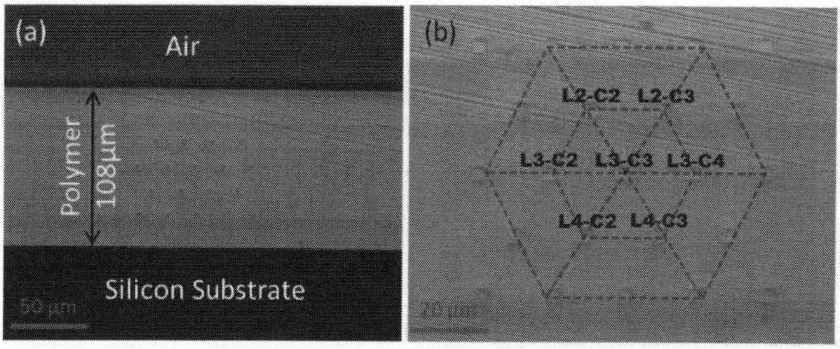

Figure 1. Microscope images of the five-layer multi-core polymer waveguides, illuminated through the cross-section. (**a**) Broader view showing the total thickness of the polymer layers on silicon substrate and (**b**) detailed view showing the alignment of the waveguides on the 20-μm hexagonal lattice.

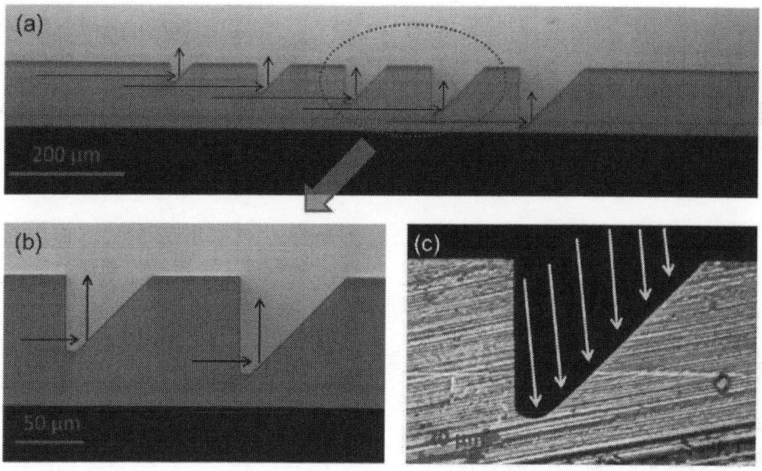

Figure 2. Microscope images of the five-step 45° mirrors. (**a**) Broader view and (**b**) detailed view showing the placement of the mirrors with backside illumination through the polymer, before the mirror surface metallization. (**c**) Detailed view illuminated from the top, revealing the metalized surface. The yellow arrows indicate the 10° angle in which the metal vapor is applied.

For testing of the mirrors, a tunable laser source is launched via a lensed fiber into the 1-cm long multicore waveguide chip and collected from the corresponding mirrors by a PD bar. The measurement setup is shown in Figure 3a. Index matching oil is used to fill the mirror groove to limit the beam divergence as well as to reduce the back reflection at the PD surface. The multi-step mirror losses are displayed in Figure 3b, ranging from 1.5 dB to 1.9 dB at 1550 nm. The bottom mirrors suffer gradually more losses, caused by the longer free-space beam propagation and associated larger beam divergence, scattering, and more material absorption in the oil filled region. The mirror loss can be further reduced by applying proper waveguide tapers to limit the beam divergence and by optimizing the dicing parameters for a smoother mirror surface.

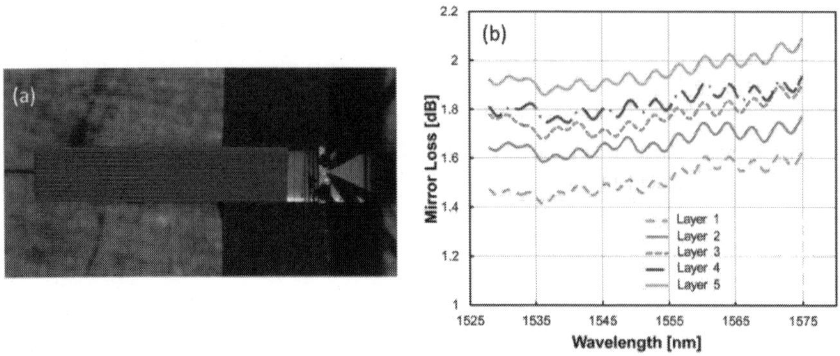

Figure 3. (**a**) Photograph of the 1-cm long multilayer waveguides and multi-step mirrors under test. Optical input (left): lensed fiber; electrical output (right): planar PD bar with probe contacts. (**b**) Mirror loss from each waveguide layer extracted from the PD responsivity.

To increase the waveguide index contrast and bring down the device footprint, amorphous silicon nitride (SiN_x) can be deposited on the polymer cladding by plasma enhanced chemical vapor deposition (PECVD) up to 300 °C [7,8,9,10]. The SiN_x / polymer heterogeneous waveguides can then be created. The aspect ratio of the SiN_x core can be adjusted to tailor the mode field diameter. For a thickness below 200 nm and a width of a few micrometers, the majority of the mode resides in the cladding and the waveguide loss depends primarily on the polymer material absorption. The minimal bending radius is 1 mm. The large aspect ratio often causes strong waveguide birefringence. Nevertheless, propagation loss below 1 dB/cm has been demonstrated for both polarizations [7].

The strong thermo-optic effect of the polymer material (coefficient ranging from $-1 \times 10^{-4}\,K^{-1}$ to $-3 \times 10^{-4}\,K^{-1}$) can be utilized for such waveguides. Gratings can be created along the SiN_x core, as shown in Figure 4a. Tunable Bragg reflectors and sampled grating filters have hence been demonstrated [7,8,9]. Further developed is a grating-assisted co-directional coupler consisting of two parallel asymmetric waveguides, one with SiN_x core and the other with polymer core [10]. The SEM photo of the coupler is shown in Figure 4b. Long-period gratings are formed along the SiN_x waveguide for phase matching and wavelength filtering. The coupler is differentially thermally tunable. The temperature gradient distribution, introduced by a pair of offset micro-heaters, can either red or blue-shift the filter curve, depending on which waveguide appears "hotter". A total tuning range of 82 nm is demonstrated experimentally.

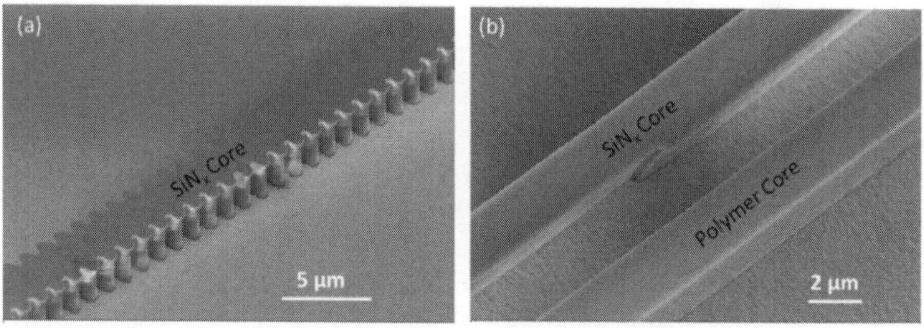

Figure 4. (**a**) SEM photo of a SiN_x waveguide Bragg grating and (**b**) SEM photo of a SiN_x/Polymer grating assisted coupler. Both photos are taken before the polymer top cladding is spin-coated.

Such waveguides have the potential to substantially enhance the versatility of polymer-based waveguide platforms. Higher contrast waveguides are feasible by using alternative materials, such as amorphous silicon with similar PECVD technology or sputtered TiO_2 and Ta_2O_5 films. Compared to the all-polymer waveguides that require at least two compatible polymeric materials with well-adjusted refractive indices, a wider choice of useful polymers can be considered, e.g., aiming at lower material absorption, larger thermo-optical coefficients and higher thermal stability.

The multi-layer and heterogeneous waveguide technology can well be combined to construct a powerful waveguide platform that can incorporate a variety of organic and inorganic materials to build, layer by layer, complex 3D PICs. High-index dielectrics can be adopted to shrink down the size of the waveguide components and the lattice can be made much more compact without severe crosstalk. Metals can be imbedded as local heaters and RF interconnects. The polymer cladding serves as an immersive motherboard, supporting the overall infrastructure and providing flexible I/Os with multi-depth fiber grooves, slots and 45° mirror arrays.

3. POLYMER AWG-BASED OLTS

Optical line terminals (OLTs) and optical network units (ONU) are basic building blocks in the WDM-PON infrastructure. The hybrid integration approach allows high-performance and cost-effective OLTs to be assembled using polymer arrayed waveguide gratings (AWGs) for wavelength multiplexing (MUX) and de-multiplexing (DMX). InP LD arrays can be direct modulated as transmitters (Tx) and planar PD arrays function as high speed receivers (Rx).

On the transmitter side, the OLT-Tx is shown in Figure 5a,b. This device consists of four InP-based 10-chanel DFB LD arrays first placed on a Si submount and then butt-joint coupled to a polymer MUX AWG. The complete device, combining the AWG and the LD arrays, is 29 mm long and 30 mm wide. The 40-channel MUX AWG is based on a polymer waveguide with square cross-section of 3.5 μm × 3.5 μm and a refractive index contrast of 1.47/1.45. It is designed to cover the L-band (1574.4 nm–1606.4 nm) with 100 GHz channel spacing. The spectral characteristic of the AWG channels is of Gaussian type with a 1-dB bandwidth of 0.25 nm. The temperature dependence is ~0.04 nm/°C. A heater electrode placed on top of the arrayed waveguide region can be biased to thermally adjust the AWG central wavelengths to match the ITU-grids [11].

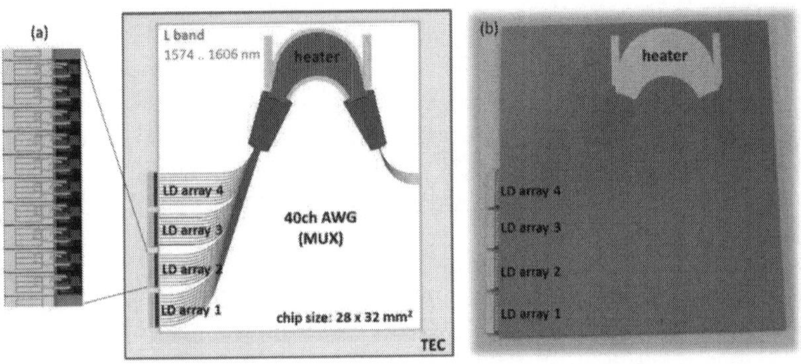

Figure 5. 40-channel OLT-Tx based on 4 × 10 direct modulated LD arrays butt-joint coupled to a thermally tunable polymer AWG. (**a**) The schematic layout and (**b**) photograph of the assembly.

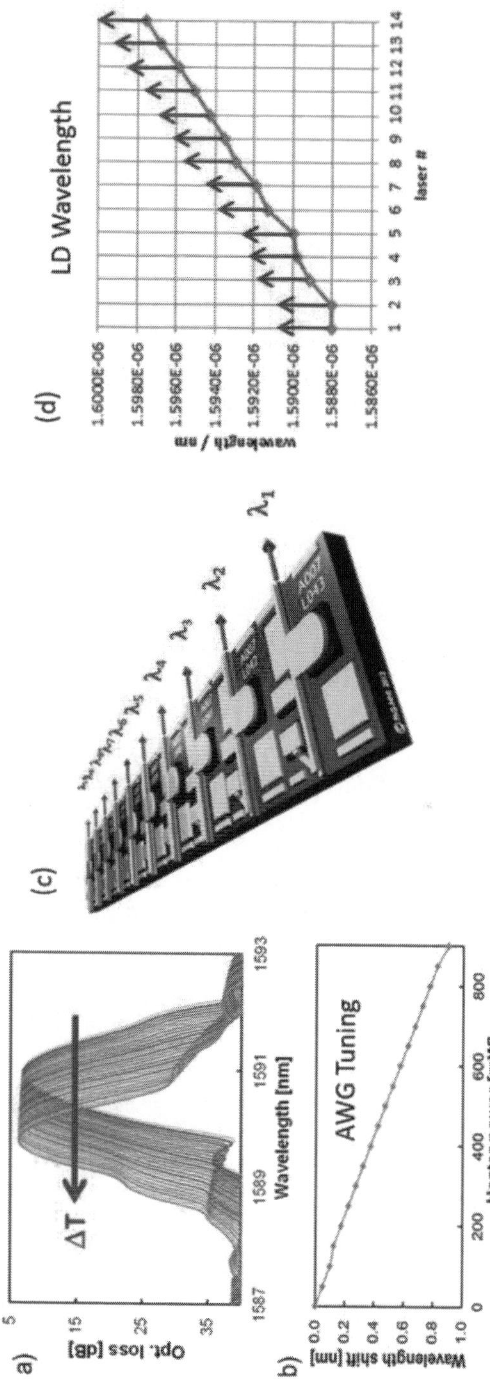

Figure 6. (**a**) Overlaid AWG spectra at different heater powers; (**b**) AWG central wavelength during thermal tuning; (**c**) schematic layout of the multi-wavelength LD array; (**d**) the measured emission wavelengths.

Figure 6a shows the overlaid spectra of the AWG at different heater powers. Figure 6b exacts and plots the central wavelength during the thermal tuning. At ~900 mW, the tuning range exceeds the channel spacing of ~0.8 nm. The schematic layout of the InP LD array is shown in Figure 6c. Each single element comprises a 150-µm-long DFB section and a 250-µm-long extra amplification section to increase the front side optical output power. The emission wavelength for each individual LD is defined by the e-beam lithography to set the 100-GHz channel spacing. A heater electrode is included in the DFB section and up to 3-nm tuning can be realized thermally, in order to pin-point the wavelengths at the ITU grid. The emitting wavelengths of different LDs on the array are listed in Figure 6d. The 4 × 10-channel LD arrays construct the 40 defined channels in the L-band. The LD arrays are first mounted on the Si-submounts with pre-defined RF lines and then butt-joint coupled to the polymer-based AWG. The arrays have been high-reflection (HR) coated on the back side against air, whereas the front facet has an anti-reflection (AR) coating against polymer to avoid back reflections. Index-matching epoxy is first applied to secure the optical coupling, followed by a mechanical fixing epoxy to support the large array.

The assembled OLT-Tx is placed on the thermo-electric cooler (TEC), which stabilizes the assembly temperature at 25 °C. The performance is evaluated and displayed in Figure 7, where the spectra of the 40 channels are measured at the outputs of the MUX AWG. An optical power larger than 0 dBm has been obtained in the fiber at a drive current ~120 mA. The spectrum of the AWG is firstly adjusted to match the central wavelengths defined by the ITU-Grid. The emitting wavelengths of the individual LDs are then aligned to the grid. The coinciding spectra from the AWG and the LDs lead to a stunningly sharp wavelength response and the achieved side mode suppression ratio (SMSR) exceeds 70 dB for all the 40 channels. With LDs (EMLs) that can be direct modulated faster than 25 Gb/s [21], a total bit rate larger than 1 Tb/s can be achieved for this type of OLT-Tx device.

The OLT-Rx at the receiver side is shown in Figure 8a,b. 4 × PD arrays, each containing 10 planar PDs are vertically coupled to the polymer DMX AWG by an integrated 45° mirror. The 40-channel polymer DMX AWG is designed to operate at 100-GHz channel spacing over the C-band (1534.2 nm–1561.4 nm). A λ/2 plate can be inserted via a deep etched slot in the middle of the arrayed waveguide region to minimize the polarization dependence [2].

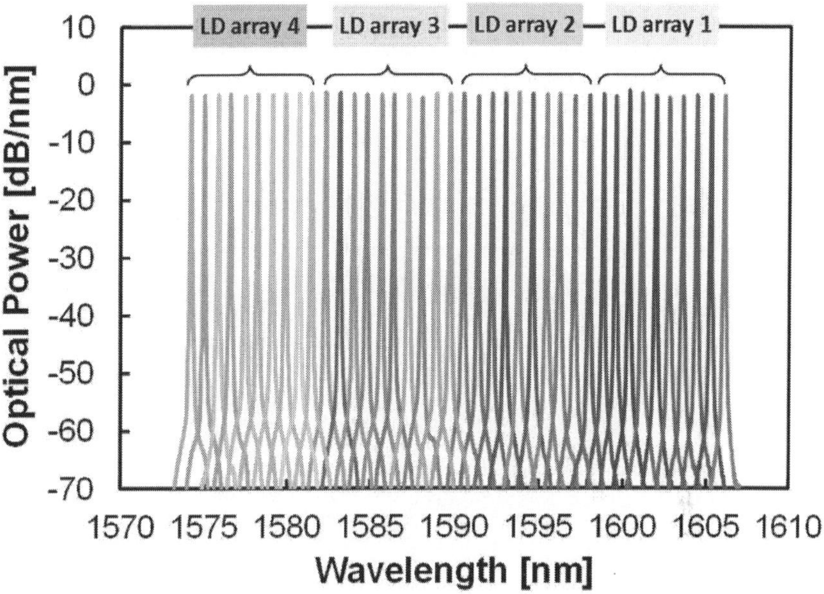

Figure 7. Multiplexed transmission spectra at the output of the OLT-Tx (measured in fiber).

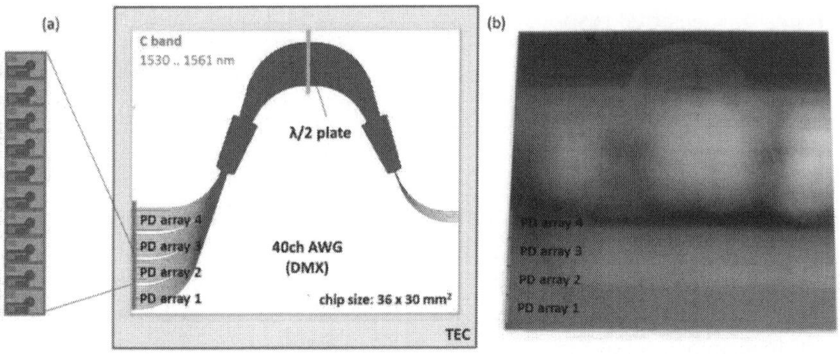

Figure 8. 40-channel OLT-Rx based on 4 × 10 planar PD arrays vertically coupled to a polymer AWG via an integrated 45° mirror. (**a**) The schematic layout and (**b**) photograph of the assembly.

Figure 9 shows the responsivities of the OLT-Rx. For the measurements, the chosen PDs have an intrinsic responsivity of 1.1 A/W and an effective responsivity of ~0.22 A/W is obtained on the OLT. The device insertion loss, including the fiber coupling loss, mirror coupling loss and the propagation loss of the polymer AWG, is calculated to be ~7.5 dB. At somewhat compromised responsivity, the planar PDs have been reported to provide a bandwidth exceeding 30 GHz. With such PDs, each channel can reach a detection rate >25 Gb/s, and a total data rate over 1 Tb/s can be realized for such an OLT-Rx.

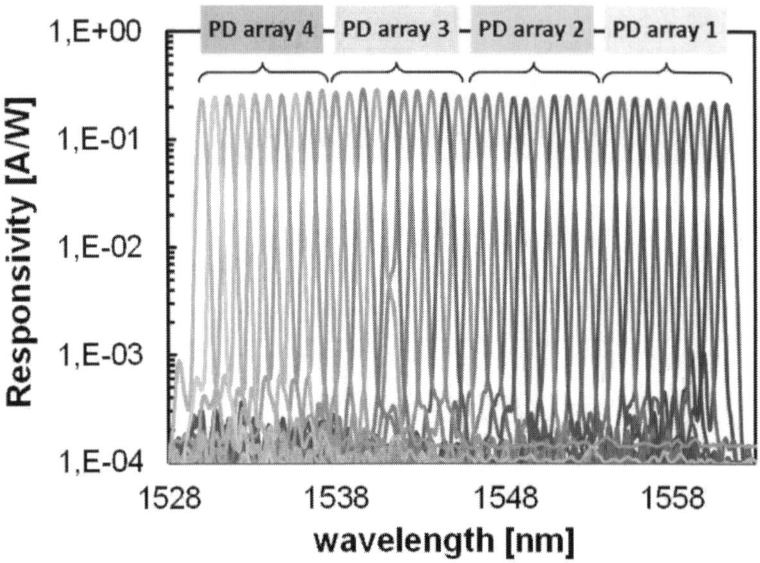

Figure 9. Measured responsivities of the PD arrays on the OLT-Rx.

4. COLORLESS ONU

Within the WDM-PON technology, a colorless optical network unit (ONU) supporting all wavelength channels is needed to simplify network operation, reduce installation cost, and keep the maintenance efforts under control [22,23]. Such a device can be assembled on the polymer platform and functions including bi-directional connectivity, real-time data rate monitoring and crosstalk suppression can be enabled simultaneously [14]. The schematic layout and the photo of the assembled ONU are shown in Figure 10a,b. The entire assembly measures 6.5 mm (L) by 2.5 mm (W).

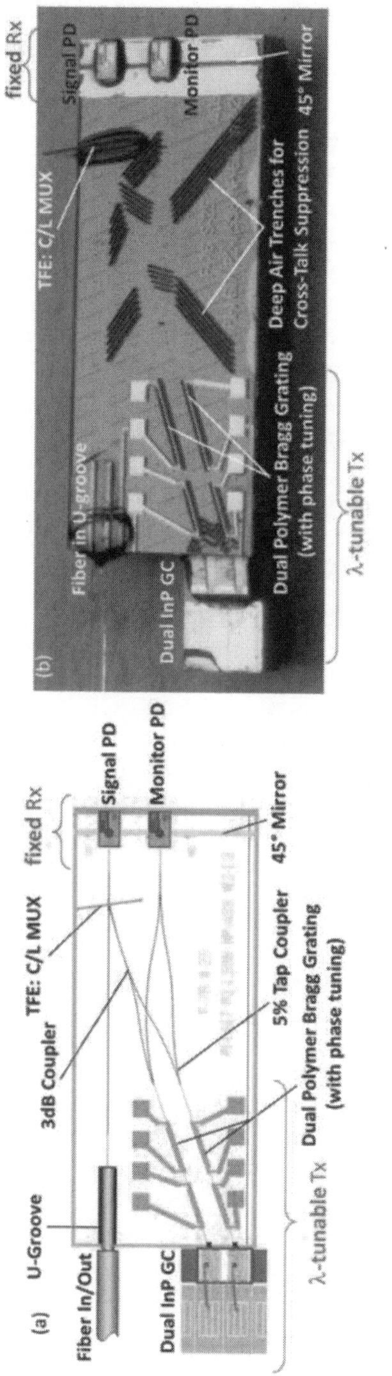

Figure 10. ONU assembled on the polymer-based hybrid integration platform. (**a**) The schematic layout and (**b**) photograph of the assembly.

The InP gain chip (GC) is similar to an LD except that the internal optical feedback loop is intentionally removed. At least one facet of the gain chip is AR coated to eliminate internal lasing and enhance light emission. It usually employs a multi-quantum well design to enhance the output power and to broaden the gain bandwidth. The dual GC array is first placed and wire-bonded onto a RF-submount and then butt-joint coupled to the dual-channel polymer Bragg gratings. Together they form a tunable transmitter (Tx) when the GC is direct modulated. This dual-channel design helps relieve the thermal strain from the single polymer Bragg grating and the wavelength tuning across the C-band is split between the two paths. In this way, the local temperature at maximal tuning position still stays far below the certified degradation temperature of 300 °C and long-term stable performance is verified over more than 6000 h at stressed conditions [24]. Five percent of the Tx power is tapped out via a pair of asymmetric Y-couplers to be received by the monitor PD. The TFE-based C/L band filter reflects the Tx signal, which further propagates through the polymer waveguide into the single mode fiber that is attached passively in the U-groove [25].

On the receiver (Rx) side, the incoming optical signal at any wavelength within the L-band goes through the same fiber into the polymer waveguide. The TFE passes the signal through to the receiver side of the ONU, where the 45° mirror deflects the light and feeds it to the high-speed planar PDs mounted on top.

A series of air trenches are etched on the polymer chip together with the U-Groove and the TFE slot. These air trenches provide many optical interfaces and deliberately scatter the stray light in the polymer slab in the direction away from the location where the Rx PD is mounted to avoid cross-talks.

The TFE is a key element performing the C/L-band wavelength (de)multiplexing. It consists of alternating layers of sputtered SiO_2 and TiO_2 films on a thin polymer substrate. The total thickness of the filter adds up to 24 µm. The fabrication is done on a standard 3-inch silicon wafer. After sputtering, the TFE is cut by an excimer laser into desired dimensions. A photo of the TFE after cutting is shown in Figure 11a. The TFE has been characterized by single mode fibers, prior to the integration into the polyboard. The reflection/transmission spectra are displayed in Figure 11b. The TFE introduces a minimal insertion loss <2 dB. Due to the band-edge effect, however, the transition slope between 1562 nm to 1572 nm causes signals at these wavelengths to suffer from extra attenuation. Filters with steeper band-edges at reasonable thickness are currently being investigated.

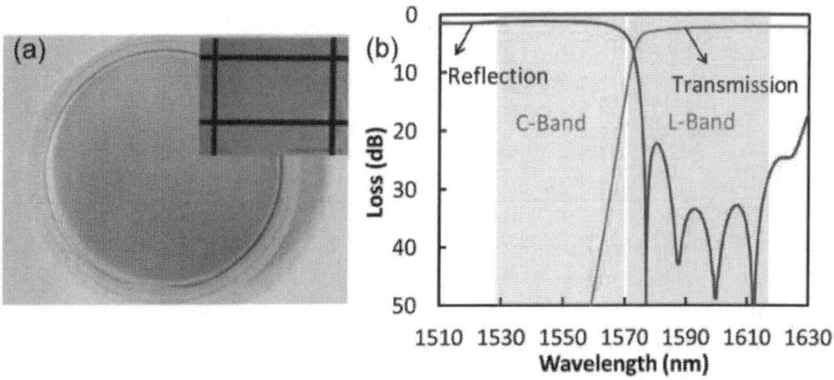

Figure 11. (**a**) Microscope image of the C/L band filter after laser cutting. (**b**) Reflection and transmission spectra of the TFE-based filter.

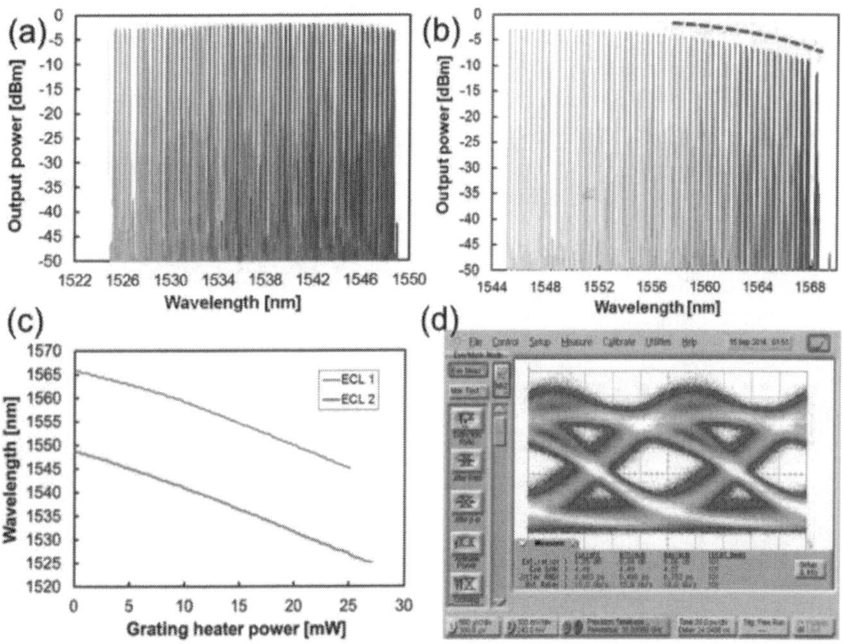

Figure 12. (**a**) Tx characteristics of the ONU. Tuning spectra of the lower arm laser; (**b**) tuning spectra of the upper arm laser; (**c**) total tuning range and efficiency of both lasers; (**d**) eye diagram at 10 Gb/s after transmission through an 11.3-km-long standard single mode fiber.

After the assembly, the ONU is first characterized on the Tx side. The drive current for the GC is set to 50 mA. The lower arm of the tunable laser path covers a tuning range from 1525 nm to 1548 nm, as indicated in Figure 12a, while the upper arm from 1545 nm to 1568 nm, as shown in Figure 12b. The side-mode suppression ratio is larger than 40 dB and the combined tuning range of 43 nm exceeds the C-band. The dropping of the laser output power at wavelength above 1560 nm is evidently affected by the band-edge of the TFE, as indicated by the dashed curve in Figure 12b. The transmitter loss away from the band-edge region is around 7 dB from the GC to the fiber, including the 3-dB intrinsic loss of the Y-branch.

Both lasers show almost identical thermal tuning behavior and the efficiencies are plotted in Figure 12c. For each laser to reach 23 nm wavelength tuning, only ~26 mW heater power is needed. The thick polymer bottom cladding and the deep-etched air trenches surrounding the buried heaters function as ideal thermal buffers [24,26,27]. The tuning efficiency reaches 0.88 nm/mW.

The Tx laser can be direct modulated on the GC at 10 Gb/s data rate [24,28,29]. The eye diagram after transmission along an 11.3-km-long standard single mode fiber (SSMF) is shown in Figure 12d. The signal to noise ratio is 5.85 dB and the bit error rate (BER) is lower than 1×10^{-13}. No band-pass filter is used in the experiment.

On the Rx side, a broadband tunable laser source is injected into the fiber and the signal PD responsivity is shown inFigure 13. Above 1580 nm, the PD responsivity remains at 0.14 A/W, corresponding to a receiver loss of ~7 dB from the fiber to the PD. The mirror loss amounts to ~1 dB. The fiber/waveguide coupling and the waveguide propagation loss add up to ~2 dB. The rest is attributed to the TFE loss, ~2 dB concluded from the measurement using fibers, and the associated scattering loss at the TFE/waveguide interface in the slot. The scattering loss is caused by the non-optimal wave-front matching from the free-space region (TFE and excess slot width) to the receiving waveguide, *i.e.*, the mismatch between the broadened beam (location and size) and the near-field mode profile of the waveguide. This loss can be reduced by introducing proper tapers and waveguide offsets at the slot region. The optical signal to noise ratio (OSNR) is measured to be 34 dB at 1580 nm and 1540 nm.

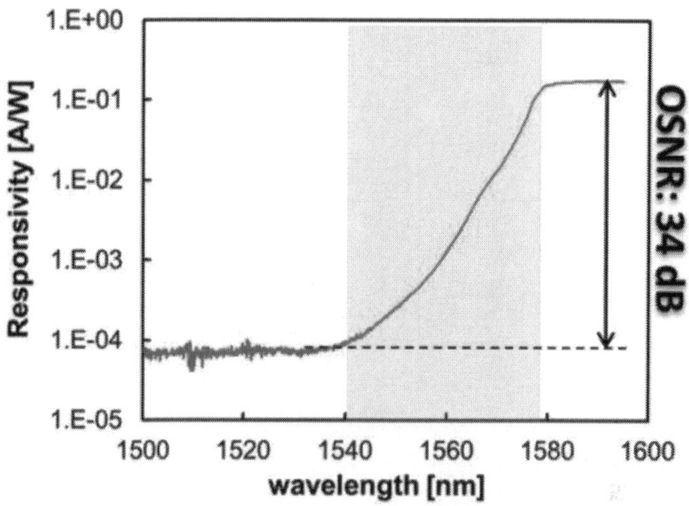

Figure 13. Responsivity measurement of the signal PD at different wavelengths.

5. DUAL-POLARIZATION COHERENT RECEIVER

The exponential growth of the internet traffic is pushing the optical fiber technology from the current 112 Gb/s links based on the DP-QPSK coherent transmission scheme to 400 Gb/s and further beyond 1 Tb/s using dual-polarization 16-QAM up to 128-QAM [30]. On the receiver side, an optical 90° hybrid is required as the frontend to convert the phase modulated signal into amplitude variations detectable by the balanced PDs [31,32]. The in-phase and quadrature components of the optical signals are then extracted as photo current variations. Further on, with the help of trans-impedance amplifiers (TIAs) and analog-to-digital converters (ADCs), the electronic signals are amplified and digitalized. Complex digital signal processing (DSP) algorithms are applied to compensate, synchronize, recover and separate the densely encoded signals.

The phase-mixing function of the optical 90° hybrid requires a laser source as local oscillator (LO). A colorless optical 90° hybrid must be able to handle a broad wavelength range, e.g., the C-band, and accordingly the LO should be tunable over that range. To bring down the device footprint, power consumption, packaging efforts and costs, an integrated, low-cost and wavelength tunable LO is highly desired.

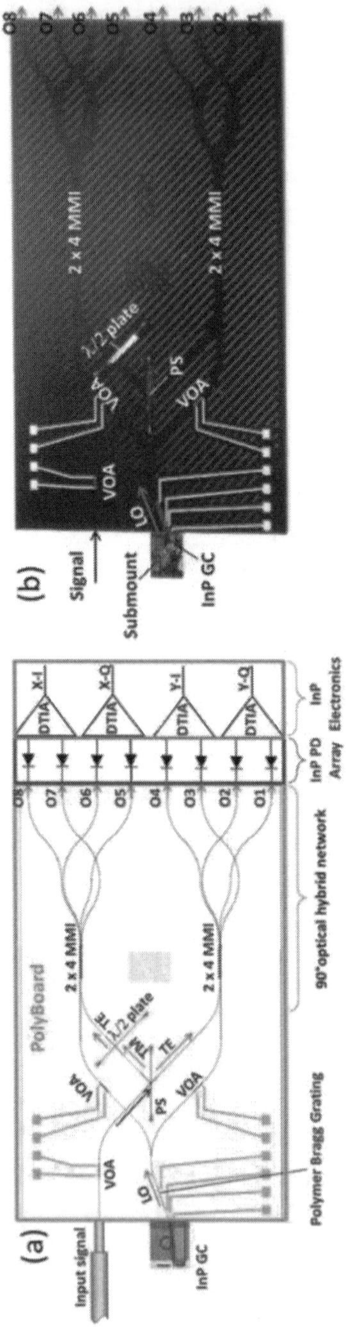

Figure 14. (**a**) Schematic layout of a coherent receiver consisting of a polymer-based 90° hybrid with integrated LO, 8 × InP PD array, and 4 × InP DTIA array. (**b**) Photograph of the assembled polymer-based 90° hybrid.

The schematic layout of the hybrid coherent receiver is displayed in Figure 14a. The device relies heavily on the hybrid integration technology: the polyboard provides low-loss waveguide components including bends, tapers and multi-mode interference devices (MMIs), as well as thermo-optic components including tunable Bragg gratings and variable optical attenuators (VOAs) [33,34]; the thin-film elements, based on the mixed layers of polymers and dielectrics, are used to split and rotate the polarization states; InP photonics platform provides the light source (GC) and the high-speed (100G) waveguide PD array: InP electronics platform offers high-speed differential trans-impedance amplifier (DTIA) and DMX circuits. The optical and electronic layouts from these components need to be arranged collectively to ensure a conflict-free assembly process. For this reason, the polymer output waveguides are extended to reach the same pitch (1 mm) as the waveguide-based PD array. The space is also reserved for a matched wire-bonding process to connect the 4 × InP-based DTIA array for the electronic signal processing.

On the polyboard, the Bragg grating is coupled to the InP GC to form the tunable LO. Three VOAs are added, both in the signal path and the LO path, to provide extra flexibility in handling signals with potentially large time- and polarization-dependent amplitude variations. The contact pads for the thermal electrodes are arranged on the top- and bottom-left side of the chip so that the electrical connections (DC) can be made in an orderly manner to the pins on the housing. Two 2 × 4 MMIs are used for the phase mixing and decoding. The thin-film-based polarization splitter (PS) separates the TE (transmitting) and TM (reflecting) components, and the half-wave ($\lambda/2$) plate rotates the TM light back to TE polarization. In this way, the MMIs and the PDs only need to be optimized for the TE polarization. It is worth noting that the PS usually causes higher loss for the transmitted TE path (~1.3 dB) than the reflected TM path (~0.5 dB). With waveguide tapers, the $\lambda/2$ plate introduces a loss of ~1 dB on the TM path, which balances out the loss imbalance and helps to generate more uniformly distributed output powers among the two mixer MMIs.

The assembled colorless dual-polarization 90° hybrid as the coherent receiver frontend has been demonstrated [35]. A photo of the assembly is shown in Figure 14b. The characterization results are summarized in Figure 15. The P-I and V-I characteristics of the LO are shown in Figure 15a, measured in the fiber at O1. The initial wavelength is 1566 nm. The output power reaches beyond 1 mW at ~400 mA gain current. Higher output power can be achieved by using InP GCs with prolonged gain region. The ripples in the P-I curve is caused by the mode-hopping and can be compensated by adjusting the phase-electrode included in the laser cavity. The LO can be continuously tuned over 22 nm, as demonstrated in Figure 15b. The side mode suppression ratio is larger than 40 dB. The power imbalances among the eight outputs are evaluated at three selected wavelengths, as summarized in Figure 15c. The power fluctuations among the wavelengths agree well with the 2 × 4 MMI measurements from the signal path and can be minimized by adjusting the MMI structures.

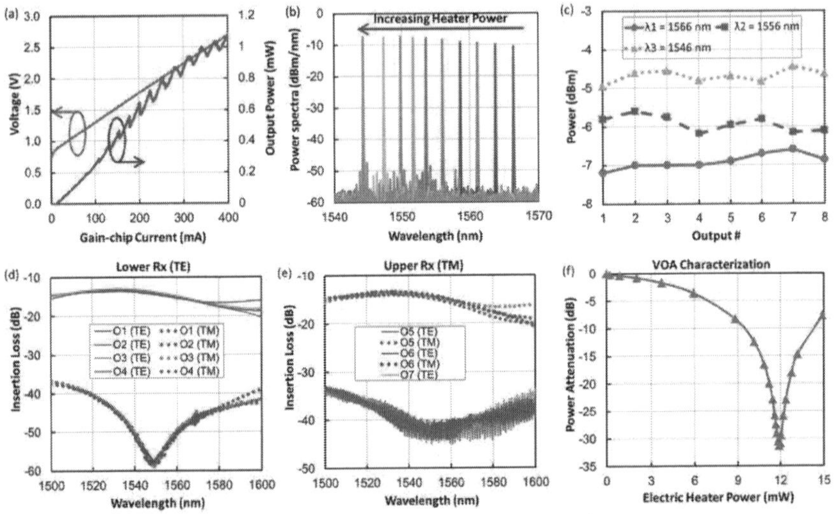

Figure 15. Characterization results of the assembled 90° hybrid: (**a**) P-I and V-I curve of the LO; (**b**) overlaid spectra of the LO during tuning; (**c**) LO power comparison at the 8 outputs at three different wavelengths; (**d**) the lower Rx O1–O4 for the TE component of the signal path; (**e**) the upper Rx O5–O8 for the TM component of the signal path; (**f**) the on-chip VOA.

On the signal path, the eight outputs are evaluated by injecting a broadband light source with controlled polarization states. The results are displayed in Figure 15d,e. Both the upper and lower Rx appear to have similar performances, with an average insertion loss of ~13.3 dB at 1530 nm and 15.5 dB at 1565 nm, including 6 dB MMI intrinsic loss, 1.5 dB fiber coupling loss (both facets), 3 dB waveguide loss (propagation + bending), 1.5 dB polarization controller loss (PS + λ/2 plate), 0.5 dB VOA loss, and 0.8–3 dB MMI excess loss across the C-band (1535 nm–1575 nm). The power imbalance among all eight outputs is below 1 dB and the polarization extinction ratio is larger than 25 dB across the C-band. The characterization results of the on-chip VOAs are shown in Figure 15f. At 12 mW electric heater power, the VOA reaches an attenuation level of 32 dB.

In the first attempt to integrate both tunable LO and VOAs on an optical 90° hybrid with dual-polarization functionality, the thermo-optic components, MMIs, and polarization controllers are all functioning properly without any noticeable optical and thermal cross-talk. To improve the performance, the tuning range of the LO can be expanded, e.g. by using a dual-grating structure and a Y-branch-based thermo-optic switch [29,33,36], to cover the full C-band. The MMI structure can be adjusted to shift the minimal loss band toward the center of the LO tuning range.

To summarize the hybrid integration technology presented in Section 3, Section 4 and Section 5, the fibers and the TFEs can be inserted passively and secured by UV-epoxy in their defined grooves and slots. The attachment of the planar PDs is assisted by a fine-placer with a positioning accuracy of ±1 μm. The process can be carried out passively with the help of the alignment marks on the chip. In the case of critical coupling of high-speed PDs with small active areas, the process can be switched to active alignment with a specially designed holder that provides electrical connections. In both cases, the exact PD location is calculated from the actual mirror width and the vertical location of the buried waveguide. Fully automated and scalable processes are being tested for the passive fiber and PD attachment.

The gain chip, LD and LD array alignment is active at the moment. The optimal position is fixed once the maximal optical power is obtained at the monitored output. An automated assisting tool has been implemented for fine-alignment using six-axis stages and nano-force sensors.

The assembly process is reproducible and the hybrid technology is well proven in the demonstrated devices. In our lab, we have assembled tens of the tunable lasers and around 80% pass our internal tests with minimal performance variation.

6. WAVELENGHT TUNABLE 100G TRANSMITTER MODULE

The polyboard technology has also contributed to the establishment of an advanced hybrid integration platform incorporating: (1) EO polymers for constructing low-complexity and low-cost Mach-Zehnder modulators (MZMs) with extremely high modulation bandwidth (>100 GHz); (2) InP components for light sources, detectors, but also high-speed electronics including MUX drivers and DEMUX circuits; (3) Ceramic (AlN) RF board and interposer that interconnects and terminates the electronic signals within the package. The entry of polyboard has not only enabled the functionalities such as wavelength tuning, polarization management, and light-flow (amplitude and direction) control, but also brought along well developed passive components (AWGs, MMIs and so on) and improved the integration compatibility with InP-based active devices and passive thin-film elements.

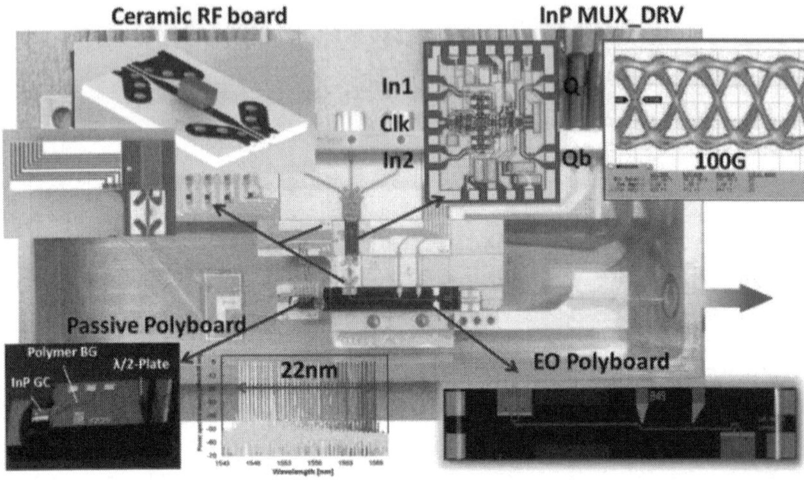

Figure 16. Dissecting of the tunable 100G optical transmitter utilizing the merits of the four material systems. From lower-right counterclockwise: EO polymers for high-speed light modulation; InP-DHBTs for ultrafast driver electronics; ceramic RF board for electronic signal interconnects and termination; polyboard for wavelength tuning and polarization control.

On this hybrid integration platform, a wavelength tunable, 100 Gbaud, all-in-one transmitter module has been demonstrated [19]. The dissecting of the module is displayed in Figure 16. The passive polyboard includes a thermally-tunable Bragg grating section and an optically coupled interface with the InP-based gain element. An external-cavity tunable laser is hereupon assembled; the half-wave plate inserted at the output waveguide efficiently converts the TE emitting laser to the required TM polarization that feeds into the EO polymer MZM.

Electro-optical (EO) polymers have been known to possess extremely fast EO responses. A modulation bandwidth larger than 100 GHz has been achieved [37,38], paving way for the advanced, complex and high-fidelity signal generations in optical communications. The polymer modulator fabrication process replies on the low-temperature thin-film-polymer on silicon (TFPS) technology [38]. The fabrication can scale up with silicon wafer size and is compatible with industry-standard semiconductor manufacturing techniques. Within this technology a poling process is required with the electric field oriented perpendicular to the wafer and the ultrafast electro-optic effect only exists for the TM polarized light. Since most LDs are developed for the TE polarization, the direct attachment of a laser source raises some challenges. The previous solution replies on the 90° rotated attachment [14,15,16]. Novel as this method is, the assembly process is complex and time-consuming. With the introduction of the polyboard, the polarization can be converted prior to the attachment. The assembly can be simplified and secured.

The electronic signal multiplexers and drivers (MUX-DRV) are fabricated on an InP-double heterojunction bipolar transistor (InP-DHBT) process. The architecture choice for this application, operating at very high bit-rate, has been to monolithically integrate a multiplexing stage with 3R function (re-amplifying, re-shaping, and re-timing). In this architecture, the electrical input interface operates at lower, half-rate speed and the 100 Gb/s electrical interfaces on chip are shortened to the minimum. The clean multiplexed signal is directly applied to amplifying stage. Two versions of InP DHBT 2:1-selector-driver have been fabricated using the 0.7 μm InP-DHBT technology [39]: with the compact lumped amplifier [40] and with distributed amplifier [41]. The differential output eye diagram of the distributed amplifier operating at 112 Gb/s has reached an amplitude of 5.9-Vpp, leading to a new record swing-speed product of 660 VGb/s.

At high speed, the electronic interconnect becomes challenging. Improper RF arrangement can throttle the overall system bandwidth to a level much below the capacity of the individual components. The RF links cannot solely rely on the bond-wires, as the dangling wires with varying structural and contact properties often lead to severe signal deterioration, interference and unbalanced transmission. An interposer is often needed to connect the electronic signals among the components and also to the cable connectors on the housing. The interposer can be fabricated by structuring metal transmission lines on a ceramic film. The design must be compatible with the housing as well as the overall placement of the individual components to avoid interference, back-reflection and other possible signal impairment. The package shown inFigure 16 is based on the gold-plated submounts of CuW that provide mechanical support for the optical and electrical components. Also included in the package are DC and RF coupling substrates, active thermal control with a thermo-electric cooler (TEC), and a lensed fiber for the coupling of the modulated signals out of the box.

The optical loss of the transmitter module is estimated to be ~12.5 dB. This includes the propagation loss on the passive polymer and the EO polymer chip, the insertion loss of the half-wave plate and the MZM, the coupling loss between the passive and the EO polymer chip and the coupling loss between the EO polymer chip and the lensed fiber. The power consumption of the device is around 7.2 W, taking into account the consumption of the tunable laser (0.25 W for the GC and max. 0.06 W for the thermal tuning), the MUX-DRV circuit (3.8 W), the phase-shifter for the biasing of the MZM (0.04 W), and the TEC (around 3 W).

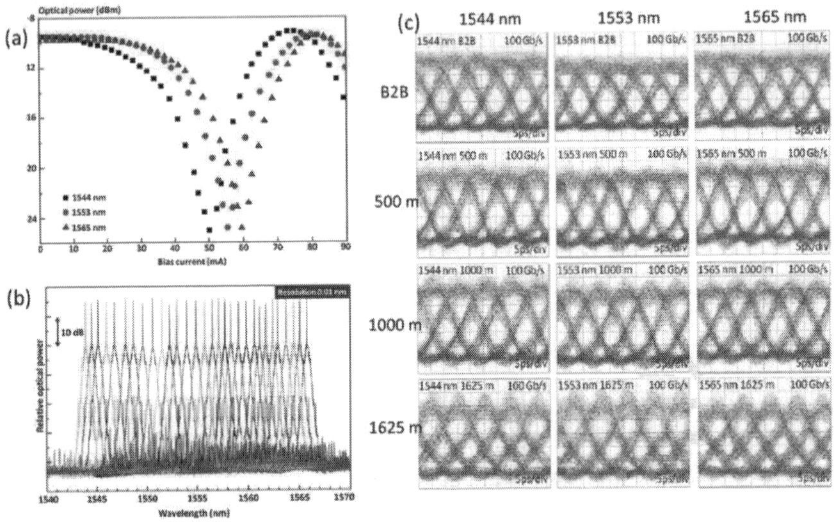

Figure 17. (**a**) The DC characteristics of the transmitter module at 1544 nm, 1533 nm and 1565 nm. (**b**) overlaid spectra of the 100 Gb/s on-off keying (OOK) signals at the output of the transmitter module covering the tuning range from 1544 nm to 1566 nm. (**c**) Eye-diagrams of the 100 Gb/s OOK signals, evaluated at three wavelengths from B2B to 1625 nm SSMF transmission.

The DC characteristics of the transmitter module, plotted in Figure 17a, are obtained by adjusting the bias current of the thermal phase shifter on one of the MZM arms. Three wavelengths are selected within the tuning range from 1544 nm to 1566 nm. The static extinction ratio is around 16 dB and the maximal output power reaches −9.5 dBm. The overlaid wavelength tuning spectra are summarized in Figure 17b, which are collected directly at the output of the transmitter when modulated with on-off keying (OOK) signals at 100 Gb/s. Continuous wavelength tuning from 1544 nm to 1566 nm has been demonstrated. The typical tones of the OOK modulation can be seen on both sides of the carriers with 100 GHz (0.8 nm) spacing.

The transmission performances at 100 Gb/s data rate are evaluated at the three selected wavelengths and the eye-diagrams are recorded for the back-to-back (B2B) transmission and propagation through a SSMF up to 1625 m. The results are listed in Figure 17c. The characteristics at the three wavelengths are very similar and the eyes remain open after 1625-m SSMF transmission. However, when the transmission distance goes beyond 1000 m, the chromatic dispersion in the SSMF starts to degrade the signal quality, indicated by the immerging hills on the top of the eye diagram. Further experiments show that error free transmission with BER lower than 10^{-10} is achievable at 100 Gb/s for a distance up to 500 m, beyond which the BER increases to 10^{-9} for the 1000-m and 10^{-7} for the 1625-m transmission, respectively. Forward error correction (FEC) coding or dispersion management in the fibers will then be needed.

The all-in-one packaged transmitter module allows direct plug-n-play deployment in the optical network systems. The single-lane 100 Gb/s connectivity can significantly reduce the number of components, shrink down device footprint and improve the network flexibility. Compared to higher-order modulation formats, the serial OOK approach is also an economical solution in terms of system complexity, number of components needed for signal generation and detection, latency, footprint, overall power consumption, and so on. The wavelength tunability can further resolve contentions, improve efficiency, and bring flexibility to the network.

To exploit this integration platform further, the MZMs and the drivers can form arrays and scale up to 400 Gb/s transmission. Combined with coherent transmission schemes and higher-order modulation formats, the transmitter can provide 1 Tb/s data rate and beyond, in which case a 3D opto-electronic integration is to be implemented, in order to distribute the electronic signals smoothly and evenly among the modulator array [3]. Besides the demonstrated wavelength tunability and polarization management, the polyboard can bring a switching matrix based on thermo-optic switches and enable the multi-flow control of the optical circuits, a function that empowers the software defined network (SDN) directly down to the optical circuit level but is deemed too expensive to be implemented on any other integration platforms. This hybrid integration platform is bound to bring new insights on the general photonic integration paradigm and serves to impact the strategic plans for the deployment of the advanced photonic devices in the telecom and datacom industry.

7. CONCLUSIONS

The polymer-based hybrid integration technology is reviewed in this work. The polymer materials can be synthesized and tested prior to the fabrication. The low-temperature layer formation process is simple and fast, leading to a substantial fabrication cost reduction. Deep grooves, trenches and slots are etched with oxygen-based RIE with a metal-mask selectivity exceeding 500. Fibers, TFEs and PDs can be passively inserted in/attached on the chip, reducing also the assembly cost.

Furthermore, multi-layer polymer waveguides and multi-step 45° mirrors enable the 3D construction of the photonic integrated circuits. Silicon nitride has been introduced as core material, and heterogeneous waveguide components are realized, opening up possibilities to incorporate inorganic materials in polymer and combine their properties for multi-functional devices.

For practical devices, 40-channel OLTs have been assembled, also proving the feasibility of integrating large arrays of LDs and PDs on the polyboard. Equipped with a thin-film filter, an on-chip fiber groove and a 45° mirror, a compact ONU device is realized, allowing bi-directional transmission in the C band and detection in the L band. For applications in coherent detection, a 90° hybrid is demonstrated with integrated LO and VOAs.

The polyboard has been further incorporated into a larger hybrid integration platform together with EO polymer modulators, InP electronics and ceramic RF interconnects. This powerful platform has yielded an all-in-one, 22-nm wavelength tunable, 100 Gbaud transmitter module that not only delivers state-of-the-art performance but also allows direct plug-and-play deployment in the optical network systems. With more advanced, multi-flow transmitter modules under way, this polymer-based hybrid integration platform is only starting to unleash its full potential and is bound to bring photonic integration to a new, unprecedented level.

Comparing to other integration platform like SOI or InP, the advantage of polymer platform lies in the low-cost fabrication technology, the possibility to realize multi-layer optical circuits, the flexibility and capability to integrate a large variety of optical functions. For applications, where several functions must be integrated on space smaller than a couple of mm^2, other integration technologies may be more appropriate.

ACKNOWLEDGEMENTS

This work was partly conducted under several projects sponsored by the German Federal Ministry of Education and Research (BMBF) including POLYBOARD and ADVANTAGEPON, and co-financed by the European Commission under the projects EU FP-7 POLYSYS (258846) and PANTHER (619411). The authors wish to thank the colleagues and the students in the Polymer OEIC group at the Fraunhofer Heinrich Hertz Institute for their excellent work in the device simulation and characterization, the Laser group for the development of InP gain chips and LDs, the detector group for providing high-speed PDs, and the TZL team for the general device fabrication.

AUTHOR CONTRIBUTIONS

Z. Zhang, D. Felipe, and N. Keil have contributed to the development of the polyboard technology. N. Keil, N. Grote and M. Schell have advised the integration concept and supervised the overall research activity. M. Moehrle has developed the InP-based laser diodes, laser arrays and gain chips. P. Runge, J.-H. Choi and H.-G Bach have developed the high speed photo detectors. D. Schmit has supervised the fabrication of the InP-based photonic components. J.-Y. Dupuy and A. Konczykowska have developed the InP-based high-speed MUX and DMX electronics. V. Katopodis, P. Groumas, Ch. Kouloumentas, and H. Avramopoulos have come up with the 100G module integration concept and performed the module characterization. G. Cangini and R. Dinu have developed the high-speed EO polymer modulators. A. Beretta, A. Dede, and A. Vannucci have designed the 100G module and performed the packaging activities. All authors have contributed to the writing, arrangement, and contents of the manuscript.

CONFLICTS OF INTEREST

The authors declare no conflict of interest.

REFERENCES

1 Keil, N.; Zawadzki, C.; Zhang, Z.; Wang, J.; Mettbach, N.; Grote, N.; Schell, M. Polymer PLC as an Optical Integration Bench. In Proceedings of Optical Fiber Communication Conference and Exposition (OFC/NFOEC), 2011 and the National Fiber Optic Engineers Conference, Los Angeles, CA, USA, 6–10 March 2011; pp. 1–3.

2 Zhang, Z.; Mettbach, N.; Zawadzki, C.; Wang, J.; Schmidt, D.; Brinker, W.; Grote, N.; Schell, M.; Keil, N. Polymer-based photonic

toolbox: Passive components, hybrid integration and polarisation control. *IET Optoelectron.* **2011**, *5*, 226–232.

3 Zhang, Z.; Katopodis, V.; Groumas, P.; Konczykowska, A.; Dupuy, J.; Beretta, A.; Dede, A.; Miller, E.; Choi, J.H.; Harati, P.; *et al.* Passive and electro-optic polymer photonics and InP electronics integration. *Proc. SPIE* **2015**, *9516*.

4 Watanabe, T.; Hikita, M.; Kokubun, Y. Laminated polymer waveguide fan-out device for uncoupled multi-core fibers.*Opt. Express* **2012**, *20*, 26317–26325.

5 Zhang, Z.; Kleinert, M.; Maese-Novo, A.; Irmscher, G.; Schwartz, E.; Zawadzki, C.; Keil, N. Multicore Polymer Waveguides and Multistep 45° Mirrors for 3D Photonic Integration. *IEEE Photon. Technol. Lett.* **2014**, *26*, 1986–1989.

6 Suganuma, D.; Ishigure, T. Fan-in/out polymer optical waveguide for a multicore fiber fabricated using the Mosquito method. *Proc. SPIE* **2015**, *9366*.

7 Zhang, Z.; Liu, D.; Felipe, D.; Liu, A.; Keil, N.; Grote, N. Polymer embedded silicon nitride thermally tunable Bragg grating filters. *Appl. Phys. Lett.* **2013**, *102*.

8 Liu, D.; Zhang, Z.; Keil, N.; Grote, N. Thermally Tunable Silicon Nitride Sampled Gratings in Polymer. *IEEE Photon. Technol. Lett.* **2013**, *25*, 1734–1736.

9 Zhang, Z.; Novo, A.; Liu, D.; Keil, N.; Grote, N. Compact and tunable silicon nitride Bragg grating filters in polymer.*Opt. Commun.* **2014**, *321*, 23–27.

10 Zhang, Z.; Genrich, G.; Keil, N.; Grote, N. Widely tunable grating-assisted heterogeneous silicon nitride/polymer waveguide coupler. *Opt. Lett.* **2014**, *39*, 162–165.

11 Keil, N.; Zawadzki, C.; Zhang, Z.; Felipe, D.; Maese Novo, A.; Brinker, W.; Klein, H.; Wagner, C.; Soares, F.; Möhrle, M.; *et al.* Progress in polymer-based components for next-generation PON applications. In Proceedings of the Optical Fiber Communication Conference (OFC 2013), Anaheim, CA, USA, 17–21 March 2013.

12 Felipe, D.; Zawadzki, C.; Zhang, Z.; Mase Novo, A.; Wenzel, M.; Li, H.; Przyrembel, G.; Sigmund, A.; Möhrle, M.; Keil, N.; *et al.* Hybrid InP/Polymer Optical Line Terminals for 40-Channel 100-GHz spectrum-sliced WDM-PON. In Proceedings of the 39th European Conference and Exhibition on Optical Communication (ECOC 2013), London, UK, 22–26 September 2013; pp. 237–239.

13 Zhang, Z.; Felipe, D.; Brinker, W.; Kleinert, M.; Maese-Novo, A.; Moehrle, M.; Zawadzki, C.; Keil, N. C/L-Band Colorless ONU Based on Polymer Bidirectional Optical Subassembly. *IEEE J. Lightwave Technol.* **2015**, *33*, 1230–1234.

14 Katopodis, V.; Kouloumentas, C.; Konczykowska, A.; Jorge, F.; Groumas, P.; Zhang, Z.; Beretta, A.; Dede, A.; Dupuy, J.; Nodjiadjim, V.; *et al.* Serial 100 Gb/s connectivity based on polymer photonics and InP-DHBT electronics. *Opt. Express* **2012**, *20*, 28538–28543.

15 Katopodis, V.; Groumas, P.; Zhang, Z.; Dupuy, J.Y.; Miller, E.; Beretta, A.; Gounaridis, L.; Choi, J.H.; Pech, D.; Jorge, F.; *et al.* 2x100 Gb/s NRZ-OOK Integrated Transmitter for Intra Data Center Connectivity. *IEEE Photon. Technol. Lett.***2014**, *26*, 2078–2081.

16 Groumas, P.; Katopodis, V.; Choi, J.H.; Bach, H.G.; Dupuy, J.Y.; Konczykowska, A.; Zhang, Z.; Harati, P.; Miller, E.; Beretta, A.; *et al.* Multi-100 GbE and 400 GbE Interfaces for Intra-Data Center Networks Based on Arrayed Transceivers with Serial 100 Gb/s Operation. *IEEE J. Lightwave Technol.* **2015**, *33*, 943–954.

17 Katopodis, V.; Zhang, Z.; Groumas, P.; Konczykowska, A.; Dupuy, J.Y.; Beretta, A.; Dede, A.; Miller, E.; Choi, J.H.; Harati, P.; *et al.* Tunable Transmitter for Serial 100 Gb/s Connectivity Inside Flexible Data Centers Using Hybrid Polymer Integration. In Proceedings of the Optical Fiber Communication Conference, OSA Technical Digest (Optical Society of America, 2015), Los Angeles, CA, USA, 22–26 March 2015.

18 Katopodis, V.; Zhang, Z.; Groumas, P.; Konczykowska, A.; Dupuy, J.Y.; Beretta, A.; Dede, A.; Miller, E.; Choi, J.H.; Harati, P.; *et al.* Tunable 100 Gbaud Transmitter Based on Hybrid Polymer-to-Polymer Integration for Flexible Optical Interconnects. *IEEE J. Lightwave Technol.* **2015**.

19 Katopodis, V.; Groumas, P.; Zhang, Z.; Dinu, R.; Miller, E.; Konczykowska, A.; Dupuy, J.Y.; Beretta, A.; Dede, A.; Choi, J.H.; *et al.* Polymer enabled 100 Gbaud connectivity for datacom applications. *Opt. Comm.* **2015**.

20 Ma, H.; Jen, A.K.Y.; Dalton, L.R. Polymer-based optical waveguides: materials, processing, and devices. *Adv. Mater.***2002**, *14*, 1339–1365.

21 Fukano, H.; Akage, Y.; Kawaguchi, Y.; Suzaki, Y.; Kishi, K.; Yamanaka, T.; Kondo, Y.; Yasaka, H. Low chirp operation of 40 Gbit/s electroabsorption modulator integrated DFB laser module with low driving voltage. *IEEE J. Sel. Top. Quantum Electron.* **2007**, *13*, 1129–1134.

22 Urata, R.; Lam, C.; Liu, H.; Johnson, C. High Performance, Low Cost, Colorless ONU for WDM-PON. In Proceedings of the National Fiber Optic Engineers Conference, OSA Technical Digest (Optical Society of America, 2012), Los Angeles, CA, USA, 4–8 March 2012.

23 Banerjee, A.; Park, Y.; Clarke, F.; Song, H.; Yang, S.; Kramer, G.; Kim, K.; Mukherjee, B. Wavelength-division-multiplexed passive optical network (WDM-PON) technologies for broadband access: A review. *J. Opt. Netw.* **2005**, *4*, 737–758.

24 Felipe, D.; Zhang, Z.; Brinker, W.; Kleinert, M.; Maese-Novo, A.; Zawadzki, C.; Moehrle, M.; Keil, N. Polymer-based external cavity lasers: Tuning efficiency, reliability and polarization diversity. *IEEE Photon. Technol. Lett.* **2014**, *26*, 1391–1394.

25 Keil, N.; Zhang, Z.; Zawadzki, C.; Wagner, C.; Scheibe, A.; Ehlers, H.; Ristau, D.; Wang, J.; Brinker, W.; Grote, N. Ultra low-loss 1 × 2 multiplexer using thin-film filters on polymer integration platform. *Electron. Lett.* **2009**, *45*, 1167–1168.

26 Liu, A.; Zhang, Z.; Liu, D.; Keil, N.; Grote, N. Thermo-optic simulations of silicon nitride/polymer hybrid waveguides.*Proc. SPIE* **2013**, *8781*.

27 Liu, A.; Zhang, Z.; Felipe, D.; Keil, N.; Grote, N. Power-Efficient Thermo-Optic Tunable Filters Based on Polymeric Waveguide Bragg Gratings. *IEEE Photon. Technol. Lett.* **2014**, *26*, 313–315.

28 Klein, H.; Wagner, C.; Brinker, W.; Soares, F.; Felipe, D.; Zhang, Z.; Zawadzki, C.; Keil, N.; Möhrle, M. Hybrid InP-polymer 30 nm tunable DBR laser for 10 Gbit/s direct modulation in the C-Band. In Proceedings of the Indium Phosphide and Related Materials Conference (IPRM 2012), Santa Barbara, CA, USA, 27-30 August 2012; pp. 20–21.

29 Felipe, D.; Zawadzki, C.; Zhang, Z.; Brinker, W.; Klein, H.; Mohrle, M.; Keil, N.; Grote, N.; Schell, M. 40 nm Tuneable Source for Colourless ONUs based on Dual Hybridly Integrated Polymer Waveguide Grating Lasers. In Proceedings of the 39th European Conference and Exhibition on Optical Communication (ECOC 2013), London, UK, 22-26 September 2013; pp. 189–191.

30 Takara, H.; Yonenaga, K.; Jinno, M. Spectrally-efficient elastic optical path networks toward 1 Tbps era. In Proceedings of the Optical Fiber Communication Conference, OSA Technical Digest (Optical Society of America, 2012), Los Angeles, CA, USA, 4–8 March 2012.

31 Runge, P.; Schubert, S.; Seeger, A.; Janiak, K.; Stephan, J.; Trommer, D.; Domburg, P.; Nielsen, M.L. Monolithic InP Receiver Chip With a Variable Optical Attenuator for Colorless WDM Detection. *IEEE Photon. Technol. Lett.* **2014**, *26*, 349–351.

32 Halir, P.; Roelkens, G.; Ortega-Moñux, A.; Wangüemert-Pérez, J.G. High-performance 90° hybrid based on a silicon-on-insulator multimode interference coupler. *Opt. Lett.* **2011**, *36*, 178–180.

33 Zhang, Z.; Keil, N. Thermo-Optic Devices on Polymer Platform. *Opt. Commun* **2015**. in press.

34 Maese-Novo, A.; Zhang, Z.; Irmscher, G.; Polatynski, A.; Mueller, T.; de Felipe, D.; Kleinert, M.; Brinker, W.; Zawadzki, C.; Keil, N. Thermally optimized variable optical attenuators on a polymer platform. *Appl. Opt.* **2015**, *54*, 569–575.

35 Zhang, Z.; Novo, A.M.; Polatynski, A.; Mueller, T.; Irmscher, G.; Felipe, D.; Kleinert, M.; Brinker, W.; Zawadzki, C.; Keil, N. Colorless, Dual-Polarization 90° Hybrid with Integrated VOAs and Local Oscillator on Polymer Platform. In Proceedings of the Optical Fiber Communication Conference, OSA Technical Digest (online) (Optical Society of America, 2015), Los Angeles, CA, USA, 22–26 March 2015.

36 Mueller, T.; Maese-Novo, A.; Zhang, Z.; Polatynski, A.; Felipe, D.; Kleinert, M.; Brinker, W.; Zawadzki, C.; Keil, N. Switchable dual-polarization external cavity tunable laser. *Opt. Lett.* **2015**, *40*, 447–450.

37 Huang, H.; Nuccio, S.R.; Yue, Y.; Yang, J.-Y.; Ren, Y.; Wei, C.; Yu, G.; Dinu, R.; Parekh, D.; Chang-Hasnain, C.J.; Willner, A.E. Broadband Modulation Performance of 100-GHz EO Polymer MZMs. *IEEE J. Lightwave Technol.* **2012**, *30*, 3647–3652.

38 Dinu, R.; Yu, G.; Miller, E.; Chen, B.; Chen, H.; Shofman, V.; Wei, C.; Mallari, J. Small Form Factor Thin Film Polymer Modulators for Telecom Applications. In Proceedings of the Optical Fiber Communication Conference, OSA Technical Digest (Optical Society of America, 2012), Los Angeles, CA, USA, 4–8 March 2012.

39 Godin, J.; Dupuy, J.-Y.; Jorge, F.; Blache, F.; Riet, M.; Nodjiadjim, V.; Berdaguer, P.; Duval, B.; Konczykowska, A. InP DHBT Mux-Drivers for Very High Symbol Rate Optical Communications. In Proceedings of the 2014 IEEE Compound Semiconductor Integrated Circuit Symposium (CSICs), San Diego, CA, USA, 19–22 October 2014; pp. 1–4.

40 Konczykowska, A.; Dupuy, J.-Y.; Jorge, F.; Riet, M.; Moulu, J.; Godin, J. InP DHBT selector-driver with 2x 2.7 V swing for 100 Gbit/s operation. *Electron. Lett.* **2009**, *45*, 1235–1236.

41 Dupuy, J.-Y.; Konczykowska, A.; Jorge, F.; Riet, M.; Berdaguer, P.; Nodjiadjim, V.; Godin, J.; Ouslimani, A. A Large-Swing 112-Gb/s Selector-Driver Based on a Differential Distributed Amplifier in InP DHBT Technology. *IEEE Trans. Microw. Theory Tech.* **2013**, *61*, 517–524.

CHAPTER 3

Harmonic Mode-Locked Fiber Laser based on Photonic Crystal Fiber Filled with Topological Insulator Solution

Yu-Shan Chen, Pei-Guang Yan, Hao Chen, Ai-Jiang Liu and Shuang-Chen Ruan *

Shenzhen Key Laboratory of Laser Engineering, Key Laboratory of Advanced Optical Precision Manufacturing Technology of Guangdong Higher Education Institutes, College of Optoelectronic Engineering, Shenzhen University, Shenzhen, Guangdong 518060, China

ABSTRACT

We reported that the photonic crystal fiber (PCF) filled with TI:Bi_2Te_3 nanosheets solution could act as an effective saturable absorber (SA). Employing this TI-PCF SA device; we constructed an ytterbium-doped all-fiber laser oscillator and achieved the evanescent wave mode-locking operation. Due to the large cavity dispersion; the fundamental mode-locking pulse had the large full width at half maximum (FWHM) of 2.33 ns with the repetition rate of ~1.11 MHz; and the radio frequency (RF) spectrum with signal-to-noise ratio (SNR) of 61 dB. In addition; the transition dynamics from a bunched state of pulses to harmonic mode-locking (HML) was also observed; which was up to 26th order.

KEYWORDS

Topological Insulator; photonic crystal fiber; mode-locked fiber laser; evanescent wave; harmonic mode-locking

1. INTRODUCTION

In the past decade, passively mode-locked fiber lasers have been extensively investigated due to its simplicity and compactness [1,2,3]. The high repetition rates of passively mode-locked fiber lasers with subnanosecond to picosecond pulses have potential applications in many fields, including high speed optical communication, optical frequency metrology [4,5]. As an important technique to increase the pulse repetition rate, the relevant harmonic mode locking (HML) fiber lasers have been a growing research region. To date, most of the previous works on HML were related to the nonlinear polarization rotation (NPR) [6,7,8,9] and nonlinear amplifying mirrors (NALMs) technique [10]. However, mode-locked fiber lasers based on NPR and the NALM technique are intrinsically environmentally unstable. Therefore, researchers have strong motivations investigate HML in passively mode-locked fiber lasers by real saturable absorbers (SAs) [11,12,13,14,15,16], such as semiconductor saturable absorbers (SESAM), single-walled carbon nanotube (SWNT) and graphene. However, these SAs also possess some inherent drawbacks. SESAMs have a narrow broadband response and require complex fabrication process and expensive equipment. SWNTs are also restricted by their diameter-dependent response spectral range, and suffered with a relatively high non-saturable loss for broadband operation. In addition, single-layer graphene is an intrinsic ultra-broadband SA, but its modulation depth is often low (typically $<1\%$ per layer [17]). Also, grapheme powder [18] and nanoscale charcoal powder [19] were employed as nano-scale saturable absorbers effectively achieve mode-locked.

Recently, another new type of Dirac material, topological insulators (TIs), which have become one of the most interesting research topics in condensed-matter physics [20,21]. At the same time, they have been successfully utilized as SA for the realization of mode-locking [22,23,24,25,26,27,28,29,30,31,32,33,34,35,36,37,38,39,40,41]. TISA could be fabricated by the mechanical exfoliation, the molecular beam epitaxy (MBE) growth, the vapor-liquid-solid growth, chemical exfoliation, the mechanical triturating and the filmy method in polyvinyl alcohol [25,42,43,44]. In most of recent studies, the nanostructured TIs are typically deposited onto the fiber ferrule [23,28,29,45]. However, the interaction length of the nanostructured TI with light is limited due to large material losses. The SA device based on the evanescent wave interaction is another effective approach to achieve mode-locked fiber laser [22,32,46,47]. Nevertheless, the microfiber-based SAs are very frangible with the disadvantage of high insertion loss (IL) and short interaction length [27], the side-polished-fiber-based SAs exhibit polarization dependent loss due to centrosymmetric structure [48]. Photonic crystal fiber (PCF) [49], characterized with regular air holes along fiber length, has been utilized as platform of SA for evanescent wave interaction by filling the nanoparticles into cladding holes [50,51,52]. More recently, the HML erbium-doped fiber lasers based on TISA has also been investigated experimentally [27,53]. However, no previous experiments have reported the HML in an ytterbium-doped all-fiber laser oscillator in the all-normal dispersion regime based on TI-filled PCF (TI-PCF) SA. In addition, it has been demonstrated that introducing proper high nonlinear effect into the laser cavity was favorable for generating harmonic mode locking (HML) pulses [27]. The small-core PCF usually has a higher nonlinearity than conventional single mode fiber (SMF). Meanwhile, TIs possess giant third order nonlinear optical property (~10–14 m^2/W) [54]. Therefore, the TI-PCF type of saturable absorber can introduce larger nonlinearity, benefiting for the generation of HML pulses.

In this paper, we reported that the PCF filled with TI:Bi_2Te_3 nanosheets solution could act as an effective SA. Employing this TI-PCF SA device, we constructed an ytterbium-doped all-fiber laser oscillator and achieved the evanescent wave mode-locking operation. Fundamental mode-locking and harmonic mode-locking (HML) up to 26th order were demonstrated.

2. SAMPLE PREPARATION AND EXPERIMENTAL SETUP

The scanning electron micrographs (SEM) of the PCF were shown in Figure 1. The total outer diameter was measured about 127 μm. The central core, fabricated by replacing the central air hole with a silica rod, had a diameter of 9.5 μm. The diameter and pitch of air holes were 5.62 μm and 8.3 μm, respectively. The fundamental and 2nd mode distribution were calculated and shown in the insets of Figure 1b. The guided light in fundamental mode could interact with the TI:Bi_2Te_3nanosheets in solution through the evanescent wave. Acting as the platform of SA, it should be pointed out that the PCF could eliminate the higher order mode in laser cavity under the following conditions: (1) the PCF was deliberately bent in our experiment tightly to suppress the high order modes; (2) Both ends of the PCF was spliced with SMF-28 fiber to avoid the excitation of high order mode. The mode field diameter (MFD) of the PCF was measured

~8.7 μm at 1310 nm. And its attenuation and dispersion parameter at 1060 nm were measured to be 2.2 dB/km and –20.5 ps/km/nm, respectively.

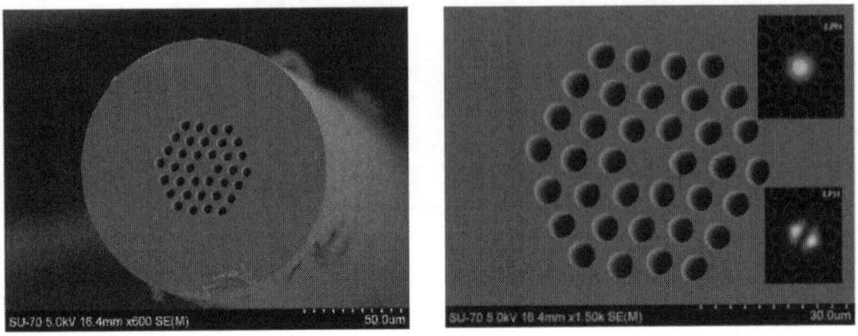

(a) (b)

Figure 1. The structure of PCF. (**a**) the whole cross section, (**b**) the enlarged microstructured region. The insets in (**b**) were calculated mode distribution of fundamental mode and 2nd mode.

The TI:Bi_2Te_3 nanosheets were synthesized with a low cost and convenient approach based on the hydrothermal intercalation/exfoliation method. Figure 2a showed the SEM of TI:Bi_2Te_3 nanosheets. The TI solution was shown in the inset of Figure 2b, which fabricated by dispersing 0.25 mg samples into 4 mL deionized water, and homogenized with 30-minites ultrasonication. The transmission spectrum of the solution was measured from 1000 nm to 1700 nm using an optical spectrometer (Perkinelmer Lambda 7500), and the linear-transmission ratio was ~43% at 1060 nm, as shown in Figure 2b.

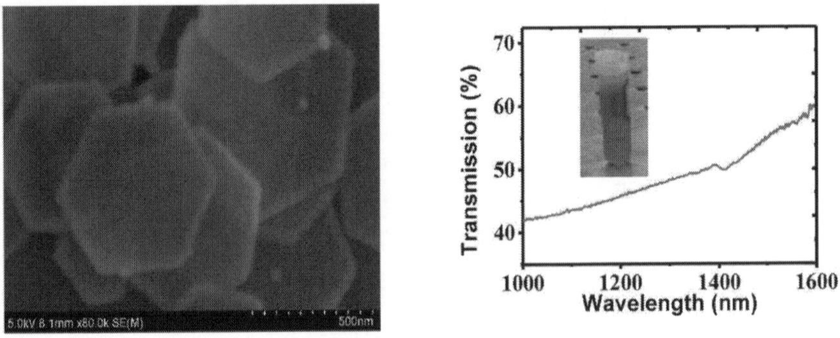

(a) (b)

Figure 2. (**a**) SEM of TI:Bi_2Te_3 nanosheets; (**b**) The transmission spectrum of the solution.

The TI solution was pulled into the air holes of PCF by an injector. The TI-PCF SA device had a length of 8cm with each end spliced with SMF-28 fiber. The measured insertion loss (IL) was only ~0.45 dB. To check the TI nanosheets really modulated the guided light in PCF, the saturable absorption property of the TI-PCF SA was measured and shown in Figure 3.

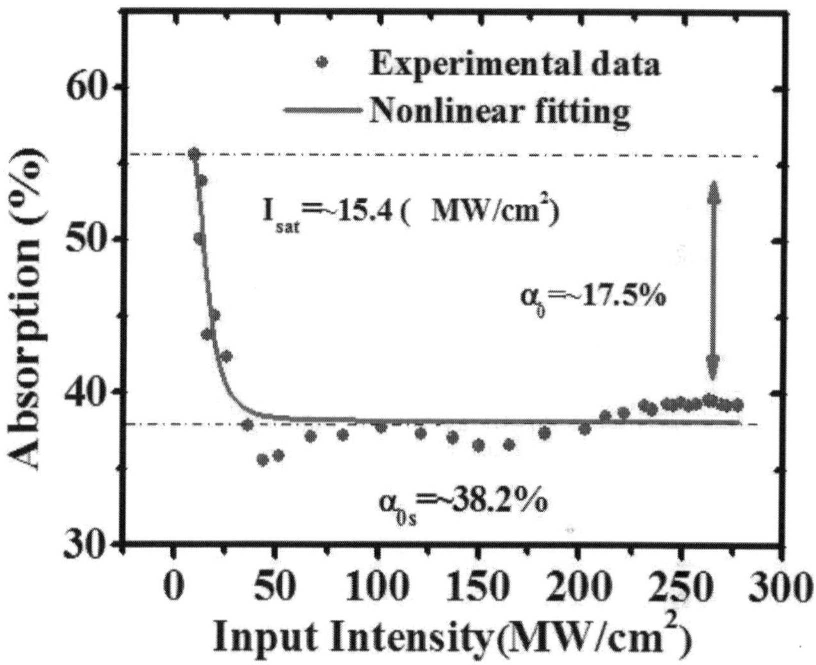

Figure 3. Measured nonlinear absorbance of the TI-PCF SA device.

The laser source used was homebuilt mode-locked oscillator with 5 MHz repetition rate and 150 ps pulse duration, which operating at ~1064 nm. The date obtained from the experiment was then fitted according to

$$\alpha(I) = \alpha_0 \left(1 + \frac{I}{I_{sat}}\right)^{-1} + \alpha_{ns} \tag{1}$$

where I was the input laser intensity, I_{sat} was the saturable intensity, $\alpha(I)$ was the intensity-dependent absorption coefficient, and α_0 and α_{ns} were the modulation depth and the non-saturable loss, respectively. The results gave a saturable intensity of ~15.4 MW/cm^2, a modulation depth of ~17.5%, and a non-saturable loss of ~38.2%. The saturable intensity of the TI-PCF SA was comparative with 53 MW/cm^2 of TI: Bi_2Se_3 deposited onto fiber end facet [24] and 12 MW/cm^2 of TI: Bi_2Se_3 embedded into PVA film [30]. On the other hand, the modulation depth of the TI-PCF SA would be high enough for stable femtosecond pulse mode-locking, considering that another TI SA with 1.7% modulation depth has demonstrated the HML according to a recent report by Z. Luo *et al.* [27].

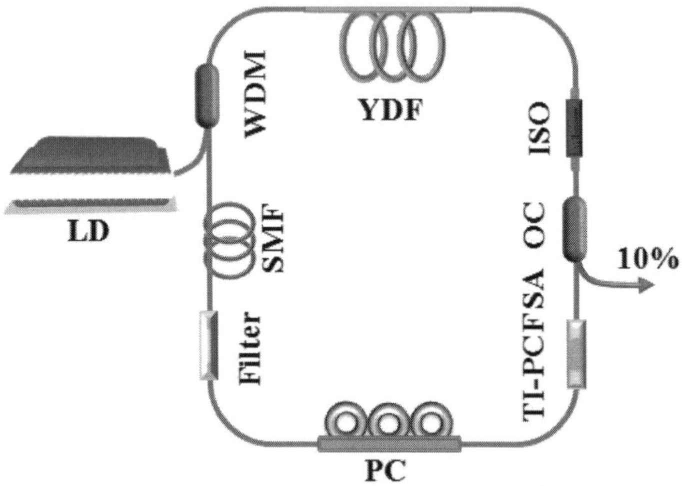

Figure 4. Schematic diagram of the fiber ring laser.

Our fiber laser was schematically shown in Figure 4. The laser was pumped by a 976 nm pump laser through a wavelength division multiplexer (WDM), and the output of the laser was coupled through a 10% fiber optical coupler (OC). A segment of ~3.4 m YDF with absorption of 250 dB/m@975 nm was used as the gain medium. A polarization independent isolator (ISO) was used to ensure the unidirectional operation of the ring laser cavity, and a three-spool polarization controller (PC) was employed to adjust the cavity birefringence. A bandwidth filter with a central wavelength of 1064 nm and 3 dB bandwidth of ~5 nm was inserted into the cavity. A total length of ~180 m HI-1060 Flex single mode fiber (SMF) was also incorporated to the cavity. The total cavity length was ~185 m. The monitoring of the output temporal pulse trains and optical spectrum were performed using a 20-GHz mixed signal oscilloscope (Tektronix MSO72004C) with a 45-GHz photo-detector (Newport 1014) and an optical spectrum analyzer (OSA, AQ6370B) with a minimum resolution of 0.02 nm, respectively.

The pump power and output power were monitored by a photodiode power meter (COHERENT). In addition, the radio frequency (RF) spectrum of the fundamental mode-locking was measured by a Mixed Domain Oscilloscope (Tektronix MDO4034B-3).

3. EXPERIMENTAL RESULTS AND DISCUSSION

3.1. Fundamental Mode-Locking

Continuous wave operation occurred at a pump power of ~76 mW. The stable operation in fundamental mode locking was observed up to the ~112 mW of pump power after careful adjustment of the PC, as shown in Figure 5a. The pulse train had a fundamental cavity frequency ~1.11 MHz, which was determined by the ~185 m cavity length. As depicted in the inset of Figure 5a, the full width at half maximum (FWHM) of the single pulse was ~2.33 ns. Figure 5b showed the typical spectrum of mode-locked pulses, it was centered at ~1062.27 nm with a 3 dB bandwidth of ~0.80 nm. In addition, the typical characteristics of steep spectral edges indicated that the mode-locked pulses have been shaped to dissipative solitons, which are found exclusively in all normal-dispersion lasers, such as ytterbium-doped all-fiber laser [55]. The corresponding RF spectrum of the laser was shown Figure 5c, and it was clearly shown that the repetition rate of the pulse train was ~1.11 MHz. The electrical signal-to-noise ratio (SNR) was ~61 dB measured with 20 kHz resolution bandwidth (RBW). It was introduced more additional linear dispersion and nonlinearity into the all-normal-dispersion cavity due to the extra fiber of ~185 m, which required for the extreme pulse broadening [56,57]. Thus, it was reasonable that the single pulse here was larger than 2 ns.

3.2. Harmonic Mode-Locking

For a passively harmonic mode-locked laser, splitting of pulses usually occurred before the attainment of nearly equal-spaced and self-organized pulse train. By increasing the pump power and adjusting the orientation of PC carefully, splitting of pulse occurred owing to the cavity pulse peak clamping effect. This process was similar to the soliton operation state in the fiber lasers with net negative cavity group-velocity dispersion [58]. The multiple pulses then rearranged themselves automatically, and all pulses were equally spaced in the cavity. Firstly, 2nd-order HML was observed at the pump power of ~127 mW by changing the PC, as shown in Figure 6a. The corresponding RF spectrum was measured, which was given inFigure 6e. It could be seen that the SNR of 2nd order HML was ~

45 dB, and its repetition rate was ~2.21 MHz (RBW was 2.50 kHz). The harmonic number could also be enlarged only by increasing the pump power slightly in our experiment, the pulse trains of the 3rd (Figure 6b), 4th (Figure 6c) and 5th (Figure 6d) order HML could be obtained under the pump powers of ~131 mW, ~140 mW, ~149 mW, respectively. In addition, the corresponding RF spectra of the 3rd (the repetition rate of ~3.32 MHz), 4th (the repetition rate of ~4.42 MHz) and 5th (the repetition rate of ~5.53 MHz) order HML were given inFigure 6f–h with the corresponding SNRs of ~ 43 dB, ~ 42 dB and ~ 42 dB, respectively. Finally, an inverse progress was explored. When the pump power was decreased to ~137 mW, the 4th order HML appeared again, then the 3th and 2nd order HML appeared in turn under a little below the previous pump powers due to the hysteresis phenomena of HML [59]. Similar to the previous work [60], the multi-wavelength mode-locked pulses were also observed in the experiment.

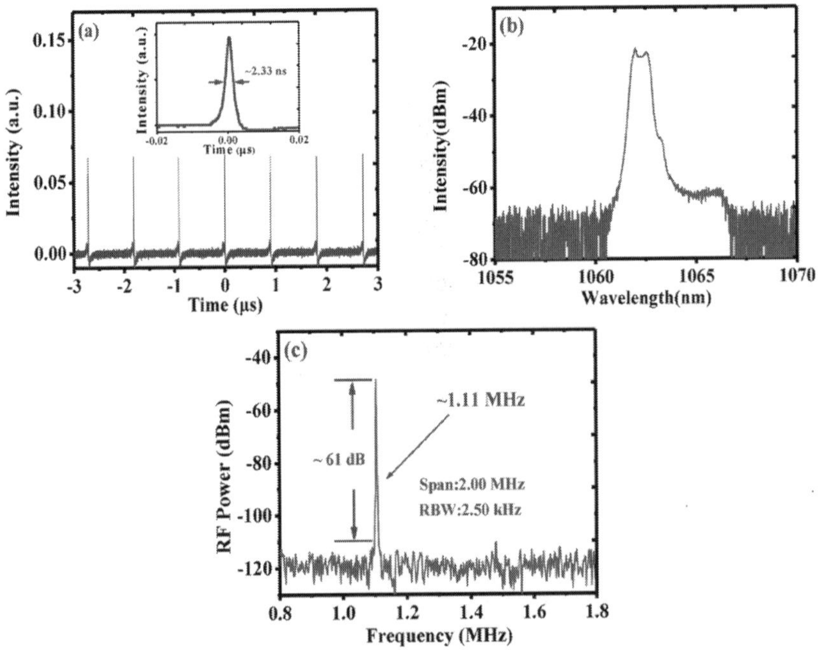

Figure 5. (**a**) Fundamental mode-locking pulse train. Inset: single pulse width. (**b**) Corresponding spectrum. (**c**) Fundamental frequency signal.

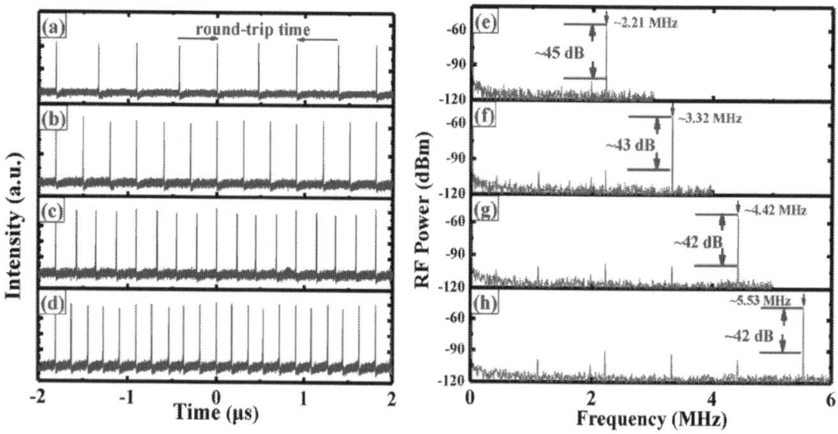

Figure 6. Pulse trains for (**a**) 2nd HML, (**b**) 3rd HML, (**c**) 4th HML and (**d**) 5th HML as well as the corresponding RF-spectra for (**e**) 2nd HML, (**f**) 3rd HML, (**g**) 4th HML and (**h**) 5th HML, respectively.

Further increasing the pump power to ~324 mW, a noticeable transition from a bunched state of pulses to high-order HML could be repeatedly achieved by tuning the orientation of PC, as illustrated in Figure 7a–e. It was clearly shown that the bunch of pulses expanded gradually and occupied the whole cavity regularly, leading to the formation of a high-order HML state eventually (shown in Figure 7e. The Figure 7f indicated that the FWHM of a single pulse in HML was ~575.80 ps.Figure 7g showed the typical output optical spectra of initial state (corresponding to the oscilloscope trace of Figure 7a and final state (corresponding to the oscilloscope trace of Figure 7e. The HML had a central wavelength of 1065.4 nm and a 3-dB spectral bandwidth of 1.51 nm. The corresponding RF spectrum was also measured, as shown in Figure 7h. The SNR of the high-order HML was ~ 38 dB, and the final repetition rate was ~28.73 MHz, corresponding to the 26th harmonic. Here we use photonic crystal fiber filled with TI:Bi_2Te_3 nanosheets solution as saturable absorber to achieve harmonic mode-locking. Another way [61] like single-wall carbon nanotube (SWCNT)-doped polyvinyl alcohol (PVA)-based as saturable absorber through using either all-physical-contact (PC) or all-angled-physical-contact (APC) connectors can also be used.

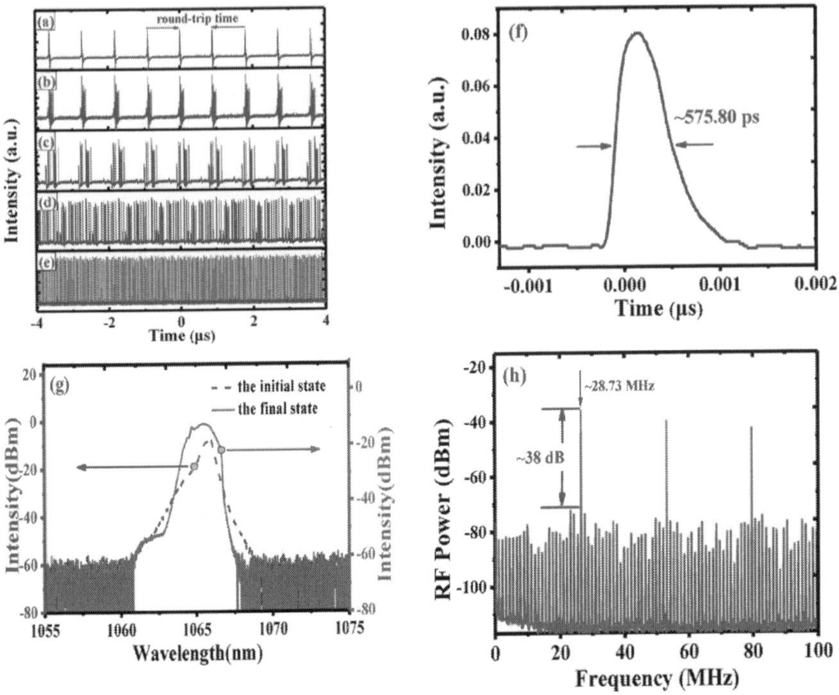

Figure 7. (**a–e**) the transition from a bunched state to the 26th-order HML. (**f**) the single pulse of 26th-order HML. (**g**) Optical spectra of initial and final state. (**h**) The corresponding RF spectrum of 26th-order HML.

4. CONCLUSIONS

As a summary, a TI-PCF SA device was firstly demonstrated, which possessed some unique advantages, such as single mode operation, nonlinearity enhancement, low insertion loss, all-fiber configuration, longer action length. Employing this SA device, an ytterbium-doped all-fiber laser was constructed and the evanescent wave mode-locking operation was achieved. Not only stable fundamental mode-locking and low order HML, but also the transition dynamics from a bunched state of pulses to high order HML were demonstrated. Our experimental results demonstrate a new scheme for ultra-fast laser mode-locking application.

ACKNOWLEDGMENTS

This research was Supported by the NSFC (61275144), Natural science fund of Guangdong province (S2013010012235), the foundation for scientific and technical innovation in Higher Education of Guangdong (2013KJCX0161), the Improvement and Development Project of Shenzhen Key Lab (ZDSY20120612094924467), the Science and technology project of Shenzhen City (JCYJ20120613172042264, JCYJ20130329142040731), Natural Science Foundation of SZU (No.201221).

AUTHOR CONTRIBUTIONS

Writing of paper: Yu-Shan Chen. Comments and feedback: Yu-Shan Chen. Experimental investigations: Pei-Guang Yan, Yu-Shan Chen, Hao Chen, Ai-Jiang Liu and Shuang-Chen Ruan. Schemes and design of investigations: Pei-Guang Yan.

CONFLICT OF INTEREST

The authors declare no conflict of interest.

REFERENCES

1 Zhang, H.; Tang, D.; Knize, R.J.; Zhao, L.; Bao, Q.; Loh, K.P. Graphene mode locked, wavelength-tunable, dissipative soliton fiber laser. *Appl. Phys. Lett.* **2010**, *96*, 111–112.

2 Wise, F.W.; Chong, A.; Renninger, W.H. High-energy femtosecond fiber lasers based on pulse propagation at normal dispersion. *Laser Photon. Rev.* **2008**, *2*, 58–73.

3 Kobtsev, S.; Kukarin, S.; Fedotov, Y. Ultra-low repetition rate mode-locked fiber laser with high-energy pulses. *Opt. Express* **2008**, *16*, 21936–21941.

4 McFerran, J.J. Echelle spectrograph calibration with a frequency comb based on a harmonically mode-locked fiber laser: A proposal. *Appl. Opt.* **2009**, *48*, 2752–2759.

5 Mikulla, B.; Leng, L.; Sears, S.; Collings, B.C.; Arend, M.; Bergman, K. Broad-Band High-Repetition-Rate Source for Spectrally Sliced WDM. *IEEE Photon. Technol. Lett.* **1999**, *11*, 418–420.

6 Zhang, Z.X.; Zhan, L.; Yang, X.X.; Luo, S.Y.; Xia, Y.X. Passive harmonically mode-locked erbium-doped fiber laser with scalable repetition rate up to 1.2 GHz. *Laser Phys. Lett.* **2007**, *4*, 592.

7 Amrani, F.; Haboucha, A.; Salhi, M.; Leblond, H.; Komarov, A.; Grelu, Ph.; Sanchez, F. Passively mode-locked erbium-doped double-clad fiber laser operating at the 322nd harmonic. *Opt. Lett.* **2009**, *34*, 2120–2122.

8 Zhu, X.J.; Wang, C.H.; Liu, S.X.; Zhang, G.J. Tunable High-Order Harmonic Mode-Locking in Yb-Doped Fiber Laser With All-Normal Dispersion. *Photon. Technol. Lett.* **2012**, *24*, 754–756.

9 Peng, J.; Zhan, L.; Luo, S.; Shen, Q. Passive Harmonic Mode-Locking of Dissipative Solitons in a Normal-Dispersion Er-Doped Fiber Laser. *J. Lightw. Technol.* **2013**, *31*, 3009–3014.

10 Chen, H.R.; Lin, K.H.; Tsai, C.Y.; Wu, H.H.; Wu, C.H.; Chen, C.H.; Chi, Y.C.; Lin, G.R.; Hsieh, W.F. 12 GHz passive harmonic mode-locking in a 1.06 μm semiconductor optical amplifier-based fiber laser with figure-eight cavity configuration. *Opt. Lett.* **2013**, *38*, 845–847.

11 Jun, C.S.; Choi, S.Y.; Rotermund, F.; Kim, B.Y.; Yeom, D. Toward higher-order passive harmonic mode-locking of a soliton fiber laser. *Opt. Lett.* **2012**, *37*, 1862–1864.

12 Jun, C.S.; Im, J.H.; Yoo, S.H.; Choi, S.Y.; Rotermund, F.; Yeom, D.; Kim, B.Y. Low noise GHz passive harmonic mode-locking of soliton fiber laser using evanescent wave interaction with carbon nanotubes. *Opt. Express* **2011**, *19*, 19775–19780.

13 Mou, C.B.; Arif, R.; Rozhin, A.; Turitsyn, S. Passively harmonic mode locked erbium doped fiber soliton laser with carbon nanotubes based saturable absorber. *Opt. Mater. Express* **2012**, *2*, 884–890.

14 Sotor, J.; Sobon, G.; Macherzynski, W.; Abramski, K.M. Harmonically mode-locked Er-doped fiber laser based on a Sb_2Te_3 topological insulator saturable absorber. *Laser Phys. Lett.* **2014**, *11*, 055102.

15 Sobon, G.; Sotor, J.; Abramski, K.M. Passive harmonic mode-locking in Er-doped fiber laser based on graphene saturable absorber with repetition rates scalable to 2.22 GHz. *Appl. Phys. Lett.* **2012**, *100*, 161109.

16 Huang, S.S.; Wang, Y.G.; Yan, P.G.; Zhang, G.L.; Zhao, J.Q.; Li, H.Q.; Lin, R.Y. High order harmonic modelocking in an allnormaldispersion Ybdoped fiber laser with a graphene oxide saturable absorber. *Laser Phys.* **2014**, *24*, 015001.

17 Martinez, A.; Sun, Z. Nanotube and graphene saturable absorbers for fibre lasers. *Nat. Photon.* **2013**, *7*, 842–845.

18 Lin, G.; Lin, Y. Directly exfoliated and imprinted graphite nano-particle saturable absorber for passive mode-locking erbium-doped fiber laser. *Laser Phys. Lett.* **2011**, *8*, 880.

19 Lin, Y.H.; Chi, Y.C.; Lin, G.R. Nanoscale charcoal powder induced saturable absorption and mode-locking of a low-gain erbium-doped fiber-ring laser. *Laser Phys. Lett.* **2013**, *10*, 055105.

20 Qi, X.L.; Zhang, S.C. Topological insulators and superconductors. *Rev. Mod. Phys.* **2011**, *83*, 1057.

21 Hasan, M.Z.; Kane, C.L. Colloquium: Topological insulators. *Rev. Mod. Phys.* **2010**, *82*, 3045.

22 Sotor, J.; Sobon, G.; Abramski, K.M. Sub-130 fs mode-locked er-doped fiber laser based on topological insulator. *Opt. Express* **2014**, *22*, 13244–13249.

23 Zhao, C.; Zhang, H.; Qi, X.; Chen, Y.; Wang, Z.; Wen, S.; Tang, D. Ultra-short pulse generation by a topological insulator based saturable absorber. *Appl. Phys. Lett.* **2012**, *101*, 211106.

24 Luo, Z.; Huang, Y.; Weng, J.; Cheng, H.; Lin, Z.; Xu, B.; Cai, Z.; Xu, H. 1.06 μm q-switched ytterbium-doped fiber laser using few-layer topological insulator Bi_2Se_3 as a saturable absorber. *Opt. Express* **2013**, *21*, 29516–29522.

25 Yu, Z.; Song, Y.; Tian, J.; Dou, Z.; Guoyu, H.; Li, K.; Li, H.; Zhang, X. High-repetition-rate q-switched fiber laser with high quality topological insulator Bi_2Se_3 film. *Opt. Express* **2014**, *22*, 11508–11515.

26 Chen, Y.; Zhao, C.; Huang, H.; Chen, S.; Tang, P.; Wang, Z.; Lu, S.; Zhang, H.; Wen, S.; Tang, D. Self-Assembled Topological Insulator: Bi_2Se_3 Membrane as a Passive Q-Switcher in an Erbium-Doped Fiber Laser. *J. Lightw. Technol.***2013**, *31*, 2857–2863.

27 Luo, Z.C.; Liu, M.; Liu, H.; Zheng, X.W.; Luo, A.P.; Zhao, C.J.; Zhang, H.; Wen, S.C.; Xu, W.C. 2 GHZ passively harmonic mode-locked fiber laser by a microfiber-based topological insulator saturable absorber. *Opt. Lett.* **2013**, *38*, 5212–5215.

28 Zhao, C.; Zou, Y.; Chen, Y.; Wang, Z.; Lu, S.; Zhang, H.; Wen, S.; Tang, D. Wavelength-tunable picosecond soliton fiber laser with topological insulator: Bi_2Se_3 as a mode locker. *Opt. Express* **2012**, *20*, 27888–27895.

29 Luo, Z.; Liu, C.; Huang, Y.; Wu, D.; Wu, J.; Xu, H.; Cai, Z.; Lin, Z.; Sun, L.; Weng, J. Topological-insulator passively q-switched double-clad fiber laser at 2 μm wavelength. *IEEE J. Sel. Top. Quant. Electron.* **2014**, *20*, 0902708.

30 Liu, H.; Zheng, X.W.; Liu, M.; Zhao, N.; Luo, A.P.; Luo, Z.C.; Xu, W.C.; Zhang, H.; Zhao, C.J.; Wen, S.C. Femtosecond pulse generation from a topological insulator mode-locked fiber laser. *Opt. Express* **2014**, *22*, 6868–6873.

31 Chen, Y.; Zhao, C.; Chen, S.; Du, J.; Tang, P.; Jiang, G.; Zhang, H.; Wen, S.; Tang, D. Large energy, wavelength widely tunable, topological insulator q-switched erbium-doped fiber laser. *IEEE J. Sel. Top. Quant. Electron.* **2014**, *20*, 315–322.

32 Lee, J.; Koo, J.; Jhon, Y.-M.; Lee, J.H. A femtosecond pulse erbium fiber laser incorporating a saturable absorber based on bulk-structured Bi_2Se_3 topological insulator. *Opt. Express* **2014**, *22*, 6165–6173.

33 Lin, Y.H.; Yang, C.Y.; Lin, S.F.; Tseng, W.H.; Bao, Q.; Wu, C.-I.; Lin, G.-R. Soliton compression of the erbium-doped fiber laser weakly started mode-locking by nanoscale p-type Bi_2Te_3 topological insulator particles. *Laser Phys. Lett.***2014**, *11*, 055107.

34 Yan, P.; Lin, R.; Ruan, S.; Liu, A.; Chen, H.; Zheng, Y.; Chen, S.; Guo, C.; Hu, J. A practical topological insulator saturable absorber for mode-locked fiber laser. *Sci. Rep.* **2015**, *5*, 8690.

35 Yan, P.; Lin, R.; Ruan, S.; Liu, A.; Chen, H. A 2.95 GHz, femtosecond passive harmonic mode-locked fiber laser based on evanescent field interaction with topological insulator film. *Opt. Express.* **2015**, *23*, 154–164.

36 Yan, P.; Lin, R.; Chen, H.; Zhang, H.; Liu, A.; Yang, H.; Ruan, S. Topological insulator solution filled in photonic crystal fiber for passive mode-locked fiber laser. *IEEE Photon. Tech. Lett.* **2015**, *27*, 264–267.

37 Huang, S.; Wang, Y.; Yan, P.; Zhao, J.; Li, H.; Lin, R. Tunable and switchable multi-wavelength dissipative soliton generation in a graphene oxide mode-locked Yb-doped fiber laser. *Opt. Express* **2014**, *22*, 11417–11426.

38 Lin, R.Y.; Wang, Y.G.; Yan, P.G.; Zhang, G.-L.; Zhao, J.-Q.; Li, H.Q.; Huang, S.S.; Cao, G.Z.; Duan, J.-A. Bright and dark square pulses generated from a graphene-oxide mode-locked ytterbium-doped fiber laser. *Photon. J. IEEE.* **2014**,*6*, 1–8.

39 Huang, S.; Wang, Y.; Yan, P.; Zhao, J.; Li, H.; Lin, R. Tunable and switchable multi-wavelength dissipative soliton generation in a graphene oxide mode-locked Yb-doped fiber laser. *Optics express* **2014**, *22*, 11417–11426.

40 Yan, P.; Liu, A.; Chen, Y.; Chen, H.; Ruan, S.; Guo, C.; Chen, S.; Li, I.; Yang, H.; Hu, J.; Cao, G. Microfiber-based WS_2-film saturable absorber for ultra-fast photonics. *Opt. Mater. Express* **2015**, *5*, 479–489.

41 Boguslawski, J.; Sotor, J.; Sobon, G.; Tarka, J.; Jagiello, J.; Macherzynski, W.; Lipinska, L.; Abramski, K.M. Mode-locked er-doped fiber laser based on liquid phase exfoliated Sb_2Te_3 topological insulator. *Laser Phys.* **2014**, *24*, 105111.

42 Bernard, F.; Zhang, H.; Gorza, S.P.; Emplit, P. Towards mode-locked fiber laser using topological insulators. In Proceedings of the Nonlinear Photonics, Colorado Springs, CO, USA, 17–21 June 2012.

43 Bansal, N.; Kim, Y.S.; Edrey, E.; Brahlek, M.; Horibe, Y.; Iida, K.; Tanimura, M.; Li, G.-H.; Feng, T.; Lee, H.-D. Epitaxial growth of topological insulator Bi_2Se_3 film on Si (111) with atomically sharp interface. *Thin Solid Film.* **2011**, *520*, 224–229.

44 Jung, M.; Lee, J.; Koo, J.; Park, J.; Song, Y.-W.; Lee, K.; Lee, S.; Lee, J.H. A femtosecond pulse fiber laser at 1935 nm using a bulk-structured Bi_2Se_3 topological insulator. *Opt. Express* **2014**, *22*, 7865–7874.

45 Chen, Y.; Wu, M.; Tang, P.; Chen, S.; Du, J.; Jiang, G.; Li, Y.; Zhao, C.; Zhang, H.; Wen, S. The formation of various multi-soliton patterns and noise-like pulse in a fiber laser passively mode-locked by a topological insulator based saturable absorber. *Laser Phys. Lett.* **2014**, *11*, 055101.

46 Zhao, N.; Liu, M.; Liu, H.; Zheng, X.-W.; Ning, Q.-Y.; Luo, A.-P.; Luo, Z.-C.; Xu, W.-C. Dual-wavelength rectangular pulse Yb-doped fiber laser using a microfiber-based graphene saturable absorber. *Opt. Express* **2014**, *22*, 10906–10913.

47 Luo, Z.; Zhou, M.; Wu, D.; Ye, C.; Weng, J.; Dong, J.; Xu, H.; Cai, Z.; Chen, L. Graphene-induced nonlinear four-wave-mixing and its application to multiwavelength q-switched rare-earth-doped fiber lasers. *J. Lightw. Technol.* **2011**, *29*, 2732–2739.

48 Choi, S.Y.; Cho, D.K.; Song, Y.-W.; Oh, K.; Kim, K.; Rotermund, F.; Yeom, D.-I. Graphene-filled hollow optical fiber saturable absorber for efficient soliton fiber laser mode-locking. *Opt. Express* **2012**, *20*, 5652–5657.

49 Knight, J.C.; Broeng, J.; Birks, T.A.; Russell, P.S.J. Photonic band gap guidance in optical fibers. *Science* **1998**, *282*, 1476–1478.

50 Liu, Z.B.; He, X.; Wang, D. Passively mode-locked fiber laser based on a hollow-core photonic crystal fiber filled with few-layered graphene oxide solution. *Opt. Lett.* **2011**, *36*, 3024–3026.

51 Lin, Y.H.; Yang, C.Y.; Liou, J.H.; Yu, C.P.; Lin, G.R. Using graphene nano-particle embedded in photonic crystal fiber for evanescent wave mode-locking of fiber laser. *Opt. Express* **2013**, *21*, 16763–16776.

52 Zhao, J.; Ruan, S.; Yan, P.; Zhang, H.; Yu, Y.; Wei, H.; Luo, J. Cladding-filled graphene in a photonic crystal fiber as a saturable absorber and its first application for ultrafast all-fiber laser. *Opt. Eng.* **2013**, *52*, 106105–106105.

53 Liu, M.; Zhao, N.; Liu, H.; Zheng, X.-W.; Luo, A.-P.; Luo, Z.-C.; Xu, W.-C.; Zhao, C.-J.; Zhang, H.; Wen, S.-C. Dual-wavelength harmonically mode-locked fiber laser with topological insulator saturable absorber. *IEEE Photon. Technol. Lett.* **2014**, *26*, 983–986.

54 Lu, S.B.; Zhao, C.J.; Zou, Y.H.; Chen, S.Q.; Chen, Y.; Li, Y.; Zhang, H.; Wen, S.C.; Tang, D.Y. Third order nonlinear optical property of Bi_2Se_3. *Opt. Express* **2013**, *21*, 2072–2082.

55 Grelu, P.; Akhmediev, N. Dissipative solitons for mode-locked lasers. *Nat. Photon.* **2012**, *6*, 84–92.

56 Kelleher, E.J.R.; Travers, J.C.; Ippen, E.P.; Sun, Z.; Ferrari, A.C.; Popov, S.V.; Taylor, J.R. Generation and direct measurement of giant chirp in a passively mode-locked laser. *Opt. Lett.* **2009**, *34*, 3526–3528.

57 Kelleher1, E.J.R.; Travers, J.C.; Sun, Z.; Rozhin, A.G.; Ferrari, A.C.; Popov, S.V.; Taylor, J.R. Nanosecond-pulse fiber lasers mode-locked with nanotubes. *Appl. Phys. Lett.* **2009**, *95*, 111108.

58 Tang, D.Y.; Zhao, L.M.; Zhao, B.; Liu, A.Q. Mechanism of multisoliton formation and soliton energy quantization in passively mode-locked fiber lasers. *Phys. Rev. A.* **2005**, *72*, 43816.

59 Liu, X. Hysteresis phenomena and multipulse formation of a dissipative system in a passively mode-locked fiber laser. *Phys. Rev. A.* **2010**, *81*, 23811.

60 Guo, B.; Yao, Y.; Tian, J.; Zhao, Y.; Liu, S.; Li, M.; Quan, M. Observation of bright-dark soliton pair in a mode-locked fiber laser with topological insulator. *IEEE Photon. Technol. Lett.* **2015**, *27*, 701–704.

61 Cheng, K.N.; Lin, Y.H.; Yamashita, S.; Lin, G.R. Harmonic order-dependent pulsewidth shortening of a passively mode-locked fiber laser with a carbon nanotube saturable absorber. *Photonics Journal, IEEE.* **2012**, *4*, 1542–1552.

CHAPTER 4

PLAT4M: Progressing Silicon Photonics in Europe

Carmelo Scarcella [1,*], Jun Su Lee [1], Cormac Eason [1], Marie Antier [2,3], Jerome Bourderionnet [2], Christian Larat [2], Eric Lallier [2], Arnaud Brignon [2], Thijs Spuesens [4], Peter Verheyen [5], Philippe Absil [5], Roel Baets [4] and Peter A. O'Brien [1]

[1] Tyndall National Institute, University College Cork, Lee Maltings, T12R5CP Cork, Ireland

[2] Thales Research & Technology, 1 Avenue Augustin Fresnel, 91767 Palaiseau Cedex, France

[3] Thales Optronique SA, 2 Avenue Gay Lussac, 78995 Elancourt Cedex, France

[4] Department of Information Technology, Ghent University-IMEC, 9000 Ghent, Belgium

[5] IMEC, 3001 Leuven, Belgium

ABSTRACT

Photonic integration is an appealing technology for emerging applications in communications, medical diagnostics and sensing. Silicon Photonics presents a highly attractive solution for large-scale photonic integration, principally because it is based on well-established CMOS-fabrication technologies. However, Silicon photonics can be difficult and expensive to implement, as it requires complex device design, fabrication and packaging capabilities. Photonic Libraries And Technology for Manufacturing (PLAT4M) is a major European project that brings together the key capabilities required to develop solutions for a range of Silicon photonic-based applications. This paper will present an overview of the PLAT4M project. It will present, in detail, a key application demonstrator (Coherent Beam Combiner), highlighting the ability of the project team to develop an integrated Silicon Photonic sub-system, from design, through to device fabrication, packaging and final test. The paper also highlights the need to consider additional capabilities besides device fabrication, such as packaging, which are critical to achieving fully operational sub-systems.

KEYWORDS

PLATM; Si photonics; coherent beam combining; Si PIC; SOI; photonic packaging; waveguide coupling

1. INTRODUCTION

The PLAT4M (*Photonic Libraries And Technology for Manufacturing*) project focuses on bringing the existing silicon photonics research platform to a level that enables transition to industry, suitable for different application fields and levels of production volume. The PLAT4M consortium includes 15 European R&D institutes and CMOS companies, key industrial and research organizations in design and packaging, as well as end users in different application fields to build the complete supply chain. Silicon has been employed for many years in CMOS integrated circuits for mass-production applications and there is the same expectation for silicon photonic technologies. Upgrading existing photonic platforms to become compatible with industrialization is mandatory at this point. It requires the establishment of design and process flows by taking into account design robustness, process variability and integration constraints. The PLAT4M partners bring a critical combination of expertise to the challenge of building a complete supply chain for commercializing silicon photonics in Europe.

The PLAT4M consortium is developing mature technologies and tools by building a coherent design flow, demonstrating manufacturability of elementary devices and process integration and developing a packaging toolkit. The project is validating the complete supply chain through application-driven test vehicles representing various application fields, such as telecom and datacom, gas sensing, LiDAR, Laser Doppler Vibrometry (LDV) and coherent beam combining. PLAT4M also focuses on preparing the next-generation platform by establishing a roadmap for performance evolution and assessing scalability to high-volume production. The supply chain is based on technology platforms of LETI (France), IMEC (Belgium) and STMicroelectronics (France and Italy), all of which is supported by a unified design environment.

In the following we present, in detail, the workflow of the silicon photonic chip for the coherent beam combining application which is a key application driver in the PLAT4M project. We present each critical stage of the development chain: motivation for the application, photonic chip design and fabrication, photonic packaging, test and characterisation.

Coherent beam combining (CBC) of fibre lasers provides an attractive mean of reaching high output laser power by scaling up the available energy while keeping fibre intrinsic advantages of compactness, reliability, efficiency, and beam quality. In CBC architectures, the power of a master oscillator (MO) is divided into N fibres that are amplified individually. The N amplified output beams are then combined coherently in order to produce ideally an optical beam with a brightness increased by N with respect to the individual beams. Moreover, this laser architecture provides a synthetic aperture with phase front control ability which permits output beam steering and atmospheric phase perturbation compensation.

The coherent addition means that the optical phases of all the channels are locked, either in a passive way [1,2] or, using active phase lock loops (PLLs) [3,4,5]. This last approach, we chose for this work, is commonly preferred when a large number of fibres are considered (typically above 10 to 20). Then, as many PLLs as the number of fibres have to be implemented, which underlines the need for collective technologies and methods. First prerequisite for a PLL is to measure the phase perturbation between the propagation channels. This can be done using various techniques [3,4,5], among which we chose an interferometric approach, which has the advantages of being collective and scalable to potentially very high number of fibres (up to hundreds) [5].

An important part of the system is the fibre channel preparation, *i.e.*, the splitting of the incident master oscillator into N channels, and the addition of one phase modulator per channel, required to feed back the system and close the PLLs. For this particular purpose, Silicon Photonics technology can bring unique advantages for these splitting and phase modulation functionalities. When a potentially very large number of channels is envisaged, volume and cost considerations become critical, and an integrated solution, carrying these functionalities on a single or few chips, is mandatory for an economically viable system. Our demonstrator aims at validating the integration of Silicon photonic functions for channel splitting and phase modulators arrays in a passive CBC experiment.

2. SI PHOTONIC IC

2.1. Design

For real system applications with Coherent Beam Combining, high optical power levels are required at the output channels of the silicon Photonic Integrated Circuit (PIC). Therefore, it is important that the total insertion loss of the PIC is as low as possible. The first aim is to increase the overall power efficiency, but it is equally important to limit the amount of total optical power required at the input of the chip. It is well known that, in silicon, the optical power that can be efficiently guided in a submicron wire waveguide is limited by two-photon absorption and subsequently free carrier absorption. At 50 mW one has to account for 3 dB/cm additional loss due to these non-linear processes. Such a power level is easily reached at the input in front of the splitter when several milliwatts of power are required at the output channels. Another crucial part in the CBC design is the phase shifter. To ensure coherent beam combining the phase shifters should be able to induce a 2π phase shift. Additionally, in order to keep the Strehl ratio degradation of the combined beam below 10%, the phase resolution should be at least $\pi/10$ and any possible spurious amplitude modulation associated with the phase modulation process should be below 1 dB [6]. Next, we will describe the design that can meet all the aforementioned requirements.

To couple light from the PM fibres in and out the silicon PIC, high efficiency grating couplers [7] are used. Higher efficiency is obtained in these grating couplers by adding and patterning a poly-silicon overlay on top of the standard crystalline silicon layer. By doing so, the directionality of the grating coupler can be enhanced, resulting in higher efficiency. The input grating coupler is followed by a 1 × 16 splitter. The optical waveguide that interconnects the grating coupler with the splitter is kept as short as possible to avoid excessive losses due to non-linear (NL) processes. The splitter itself is implemented as a 4-stage tree of 1 × 2 MMIs. MMIs are known to be very robust against process variations, having low insertion loss and high channel uniformity. A disadvantage of the tree architecture is the high power in the first branch of the tree, which is the sum of the power in all the output channels, typically 20/30 dBm. However, for a full system demonstration as shown here, this approach was selected as being the most reliable. The output channels of the splitter are connected to the phase modulators. Because carrier based silicon phase modulators inherently induce strong spurious amplitude modulation, this type of modulator cannot be used. Fortunately, because the required speed for this application is only in the kHz range, thermo-optic phase shifters offer a good alternative, as they have low insertion loss and low spurious amplitude modulation.

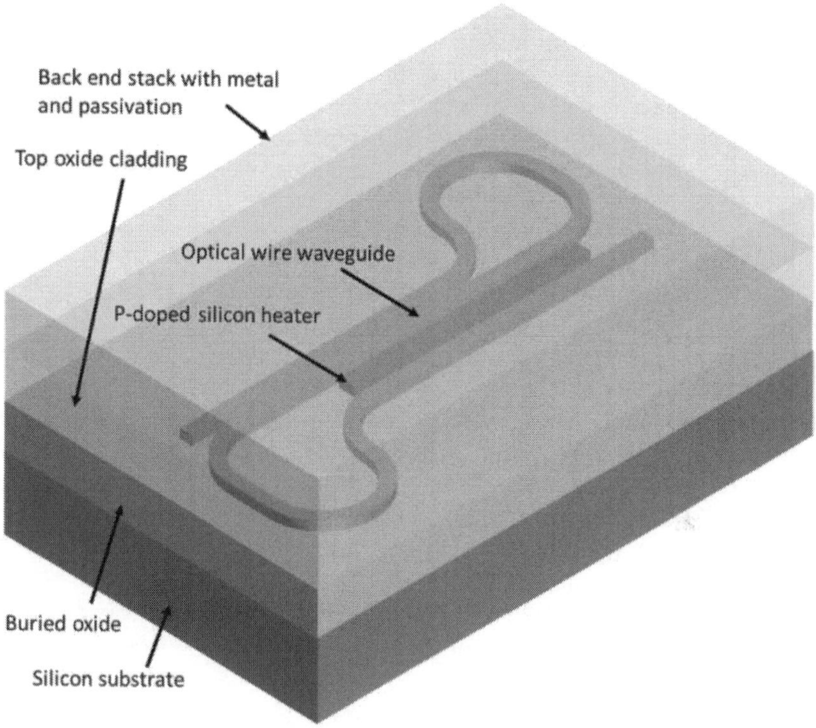

Figure 1. Schematic representation of the thermo-optic phase shifter implementation, where the optical waveguide is wrapped around a doped silicon line.

The thermo-optic phase shifters are implemented as p-doped silicon lines in the same layer as the optical waveguides. The fact that these heaters are in-plane with the optical waveguides offers an advantage over also often used metal heaters that are positioned above the optical waveguide. The reason is that in the case of in-plane heaters the optical waveguide can be wrapped around the heater, which gives an efficiency improvement of at least a factor of 2 without compromising for speed [8]. Figure 1 shows a schematic representation of the phase shifter that has been implemented in the CBC design. The optical waveguide passes by the heater as close as possible and is then routed back to pass by the other side of the heater after which it is again routed towards the direction of the output waveguide. The heater width and spacing with respect to the optical waveguide are determined from optical and thermal simulations. From a thermal point of view, it is desirable to put the heater as close as possible to the waveguide to enhance the efficiency. However, a directional coupler is formed by the optical waveguide and the silicon line heater, which limits the minimum spacing because of possible undesired coupling. To minimize the coupling, the width of the heater can be modified to avoid phase matching between the optical modes that can exist in the optical waveguide and the silicon heater line. According to our simulations a heater width of 1.2 μm is a good choice and this allows placing the heater at a distance of 600 nm from the optical waveguide without causing significant undesired coupling when the heater has a length of 200 μm.

The outputs of the thermo-optic phase shifters are connected to the high efficiency grating couplers which are arranged in an array with 127 μm pitch for interfacing with a fibre array. In order to comply with packaging requirements, shunt waveguides with grating couplers have been added around the input- and output grating couplers to enable an active alignment approach. Furthermore, optical in- and outputs are placed at the left and right side of the chip, while all electrical bondpads for electrical contacting were routed to the top of the chip for wire bonding. Because of the size of fibre array heads, the distance between the grating couplers and electrical bondpads should be large enough such that the fibre array heads do not block the electrical bondpads. Note that the size of the design is mainly determined by these packaging constraints. It is therefore important to already be aware of these constraints in the design phase of the chip, as in worst case a whole new design cycle can be required which is a significant cost.

As handling high optical powers is crucial for practical CBC systems, a new integrated grating coupler splitter has been developed in parallel with the full demonstrator [9]. This integrated grating coupler splitter consists of a star coupler that has a curved grating in its slab region. The incoming light from the fibre is then "defocused" by this curved grating into the slab of the star coupler. As the light is now never confined in a single mode waveguide, optical intensities are lower and as a result non-linear effects can be avoided. This device is capable of handling input powers up to 1 W and will be implemented in the next generation designs.

2.2. Fabrication

The devices were fabricated in IMEC's silicon photonics full platform (ISIPP25G) [10] on 200 mm SOI wafers with 220 nm silicon on a 2 μm buried oxide. First, the passive waveguide layer is processed using three etch steps: a complete 220 nm etch, a partial 70 nm etch and a partial 160 nm etch. In our design we used the deep etch for the waveguides and the doped silicon heaters while the 70 nm etch was used to lower the refractive index contrast in the lateral direction in the MMI and to define the first part of the high efficiency grating couplers. These structures are then oxide-clad and planarized, and on top of 70 nm partial etch for the grating couplers an additional 160 nm of poly-silicon is deposited. This layer is then patterned to define the high-efficiency gratings. Next, the dopants for the heaters are implanted and activated. Local silicidation is applied for contacting and after oxide cladding deposition contact holes are etched and filled with Tungsten. Finally, a standard Cu-damascene back-end and passivation is added. A microscope image of the fabricated demonstrator circuit is shown in Figure 2.

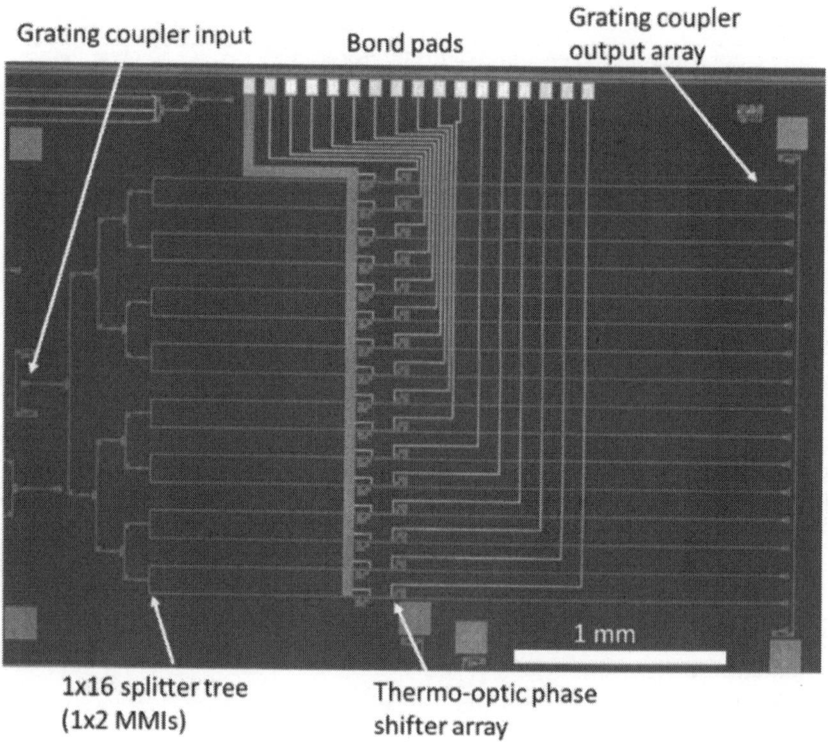

Figure 2. Microscope image of the demonstrator circuit after finalizing the silicon Photonic Integrated Circuit (PIC) fabrication process.

3. PHOTONIC PACKAGING

The photonic chip is housed onto a custom test board that allows us to connect the 16 heaters to a DC connector. A 10 kΩ thermistor is placed in good thermal contact with the IC. The printed circuit board (PCB) presents thermal vias on the PIC footprint, and a thermoelectric cooler is mounted underneath the PCB for temperature control of the device. The test-unit is assembled on top of an optical breadboard that acts also as heat-sink for the thermoelectric cooler (TEC).

The cross-section of single mode Si photonics waveguides @ 1550 nm is 220 nm × 450 nm. Diffracting grating couplers [7] are used for optical coupling between waveguides and single-mode fibres. For Coherent Beam Combining, it is necessary to use polarization maintaining (PM) fibres to connect to the Silicon waveguides. The 1D grating couplers implemented on this demonstrator support only transverse-electric (TE) polarization of light. The fibre stress rod orientation shown in Figure 3a allows only the TE mode to reach the grating couplers. Active alignment is required for optical coupling between single-mode fibres and PIC [11]. We use two extra grating couplers or “optical shunts” in addition to those on the input signal channels, Figure 3b. These shunts are necessary for accurate fibre alignment and they avoid the need to use the signal channels during the alignment process. The active alignment process involves maximizing the signal coupled into the shunt waveguide when the fibre array is scanning in a near contact position to the Si-PIC surface.

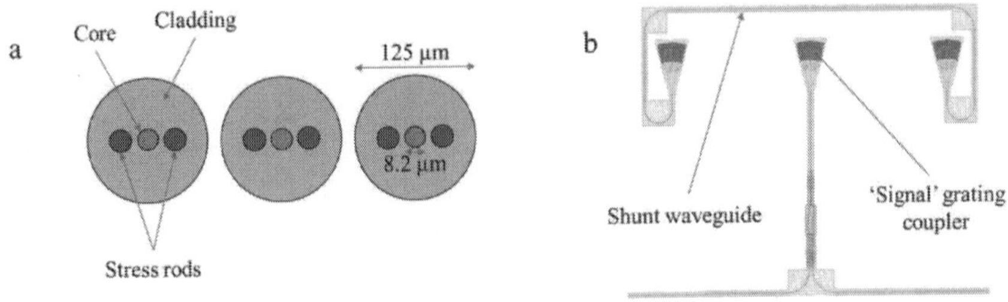

Figure 3. (**a**) Polarization maintaining fibre array with stress rods oriented to deliver transverse-electric (TE) mode on the grating couplers; (**b**) Shunt waveguide connecting two dummy grating couplers for active fibre coupling process.

Grating couplers designed for this application require light incident at 10° off-vertical and the channel pitch is 127 µm, corresponding to standard fibre arrays. When the fibre array is positioned on the maximum coupling efficiency point through the shunt waveguide, a thin layer of low-shrinkage UV cure epoxy is used to bond the fibre block onto the PIC surface, see Figure 4a. The epoxy refractive index (n) is approximately 1.5, which is close to the refractive indices of the fibre and the SiO_2 cover layer on top of the grating coupler. Figure 4b shows the 3D model of the fibre connection to the PIC. Two different fibre arrays have been used for the input and the 16 outputs. On both sides, two grating couplers shorted by a waveguide enable the active alignment of the fibres.

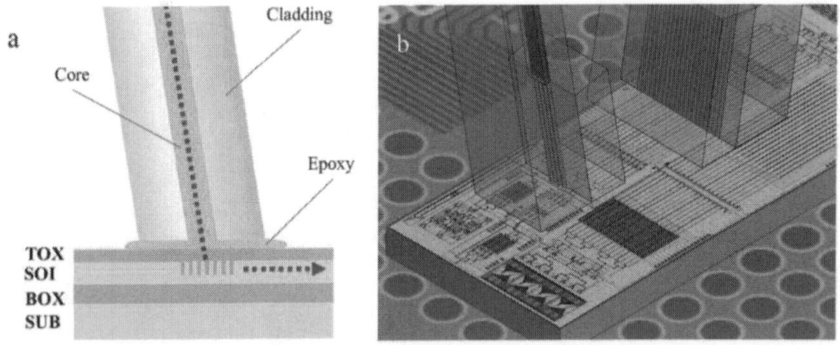

Figure 4. (**a**) Fibre connection to the grating coupler. The fibre block is polished at 10° and a thin layer of "optical" epoxy is used to bond the fibre onto the photonic chip; (**b**) 3D model of the fibre connection to the PIC. The PIC presents one input and 16 outputs. On both input and output side two additional grating couplers have been added to support the active fibre alignment.

Figure 5 shows a completed prototype, note the two posts supporting the bulky fibre bundle to avoid unnecessary stress on the joint between fibre array and PIC. It should be noted that it is possible to use a planar fibre array for a more compact overall assembly [11]. However, this initial CBC prototype used the vertical fibre design. As mentioned previously, this prototype includes 16 channels, which is still far from the hundreds or more potentially required for extremely high power CBC sources. The main challenge for scaling up this channel count is on the packaging side. With the current fibre coupling approach, the chip area increases linearly with the number of channels, with huge impact on the cost. However, 2D fibre arrays should be a valuable implementation in the future, to scale up to tens of channels on a chip with moderate dimensions.

Although the chip fabrication process is ready for large scale production, the packaging process is based on active alignment and not yet suitable for high volume manufacturing. This lack on the packaging side will be soon addressed by industry using vision based optical coupling techniques.

Figure 6a shows the wavelength response across the shunt waveguide for the four demonstrators prepared. It is peaking at 1535 nm. At 1550 nm, the CBC operation wavelength, the insertion loss for each grating coupler is 4 dB with less than 1 dB fluctuation, see Figure 6b. It represents also the insertion loss of the device in each channel (excluding the 1:16 splitter block). The grating couplers are designed to peak at 1550 nm with a "fibre-air-chip" interface when flat-end fibres at 10° off-vertical are used for optical coupling. This interface is not present in the packaged device because index matching epoxy is used to bond the fibres onto the chip. As a result there is not refraction at the silicon oxide interface and the beam shines the grating coupler with a different angle resulting in a shift of the peak wavelength.

Figure 5. Photo of the coherent beam combining (CBC) prototype. The Si-PIC is housed onto a custom printed circuit board (PCB) and two fibre arrays are aligned and attached on top of the Photonic Integrated Circuit (PIC) using ultraviolet cure epoxy.

We also investigated the alignment tolerance of our grating couplers using standard single-mode fibres (SMF-28), seeFigure 7. The coupling efficiency through the shunt waveguide was recorded while the fibre array was swept over the grating coupler with a 200 nm step. The alignment tolerances results are relatively relaxed: about 1 dB of additional losses over a planar misalignment of ± 2.5 μm, which is in good agreement with values reported in literature [12]. Figure 7a shows the 2D map of the grating coupler alignment tolerances over an area 15 μm × 15 μm. Figure 7b shows the grating coupler efficiency recorded moving the fibre along two orthogonal axes on the grating plane.

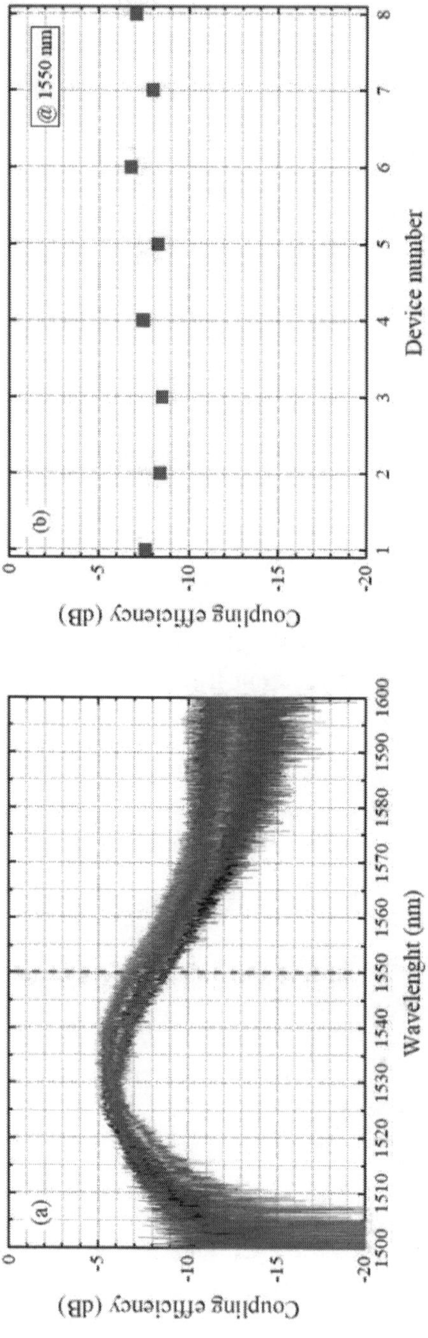

Figure 6. (**a**) Grating coupler wavelength response on the eight measured samples. The coupling efficiency peaks at 1535 nm and is still acceptable at 1550 nm; (**b**) Coupling efficiency at 1550 nm of the eight tested structures. The insertion loss on each grating coupler is of about 4 dB and the uniformity is better than 1 dB.

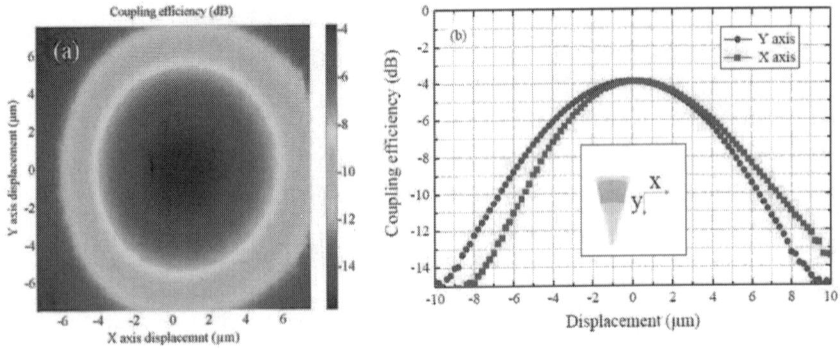

Figure 7. (**a**) Grating coupler alignment tolerance measurement; (**b**) X and Y scan of the fibre on the grating coupler plane showing about 1 dB of extra insertion losses with a fibre displacement of ± 2.5 μm.

4. EXPERIMENTAL SETUP

As described in Figure 8, the principle of this measurement is to retrieve the optical phase distribution from the spatial position of the fringe sets resulting from the interference of each individual beam with a common reference plane wave. This collective phase measurement is operated at 1 kHz refresh rate, which gives the PLL loop frequency.

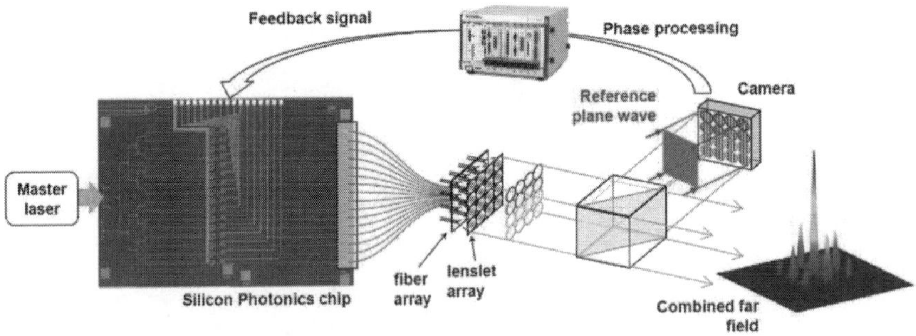

Figure 8. Schematic of the CBC experimental setup with active phase locking.

In our CBC system, a 1.55 μm CW master oscillator directly feeds the Silicon chip through a PM fibre glued onto the input grating coupler. The power of the master oscillator is split on chip into 16 channels, each of which includes a thermal phase modulator. The outputs of the 16 waveguides are collectively out-coupled from the chip using a PM optical fibre array aligned and glued onto the PIC's output grating couplers array. At the other end of fibre array, the 16 fibre outputs are arranged in a 4 by 4 squared lattice, and collimated by a microlens array to form 16 collimated and parallel beamlets. Details on the fibre array laser head can be found in [13]. The laser intensity at each of the 16 fibres output is first measured and normalized to the energy delivered by the master laser.

5. RESULTS AND DISCUSSION

The fibre-to-fibre transmission measurement is shown in Figure 9a. The overall insertion loss homogeneity is about 1 dB, and the insertion loss of the device is 7 dB per channel, excluding the 1:16 ratio of the splitter tree. The phase responses of the PIC's modulators are collectively measured using our CBC interferometric set up, by applying square shaped voltage waveforms at 1 kHz frequency to the modulators. Therefore, we make sure that the thermal modulator technology is suitable with our chosen 1 kHz loop frequency for the PLLs. As shown in Figure 9b, the phase responses of all the 16 channels are very homogeneous, with a slope efficiency of 0.285 rad/mW. Therefore, a 2π phase shift is achieved for 22 mW of electrical driving power.

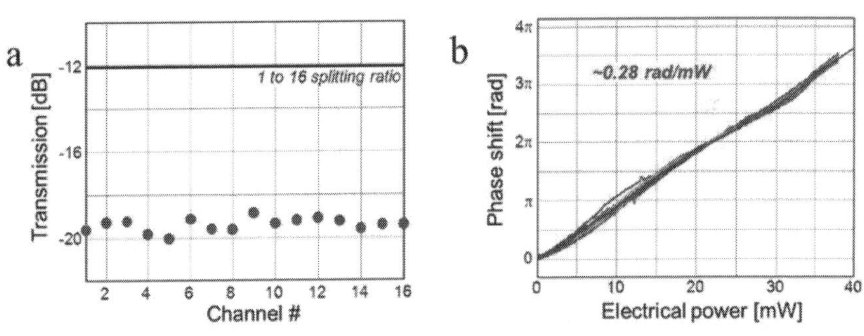

Figure 9. (**a**) Fibre-to-fibre transmission of the 16 channels; (**b**) Phase modulation response for the 16 channels (superposed colored lines).

Finally, the successful coherent combination is evidenced by recording the far field of the 16 parallel and collimated output beams. As shown in Figure 10, when the PLLs are off, the far field pattern is a fluctuating speckle-like profile resulting from the interference of the 16 beams with random phase distribution. When the PLLs are on, the phases of all the 16 channels are locked together and form an overall plane phase front for the composite output pupil of the system. This result in the far field pattern is shown in Figure 10 (right), with a bright and stable central lobe. The intensity lost here in the satellite lobes is just a result of the filling factor of the individual beams in the collimating microlenses plane. The temporal stability of the system is also illustrated in the plot below, with the energy encircled in the main central lobe plotted against time. The measured channel-to-channel residual phase error in closed loop configuration is below λ/120.

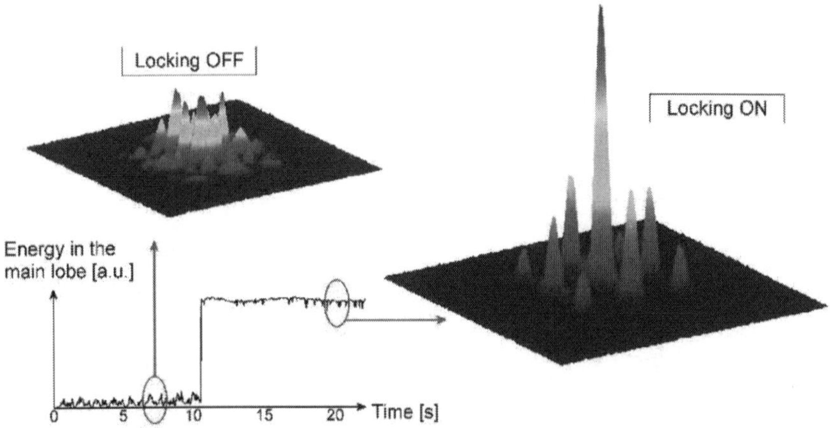

Figure 10. Experimental far field pattern of the 16 combined beams, when PLLs are off (**upper left**), and when PLLs are on (**right**). The plot shows the temporal evolution of the energy encircled in the main central lobe of the combined beam.

The optical power incident on the chip here was 10 dBm, meaning approximately −9 dBm at each fibre output. This is still too low for a real system application where fibre amplifiers have to operate in saturation regime. The next generation of devices will be able to handle one to two orders of magnitude higher input power.

6. CONCLUSIONS

The growing interest in Silicon photonics is mainly driven by its ability to satisfy demands in large markets, particularly for datacoms, medical diagnostics and sensing applications. Device fabrication based on CMOS wafer-scale processes can meet this demand. However, device fabrication is only one element of the supply chain necessary to produce fully working sub-systems. In this paper, we have shown how the PLAT4M project is addressing this technological supply chain challenge, by bringing together a range of capabilities from around Europe, from Si-PIC design, through to device fabrication, packaging and final test. We reported the demonstration of this full supply chain through one of the applications developed in PLAT4M, a Si-PIC Coherent Beam Combiner. It is however worthwhile mentioning that the PLAT4M supply chain is also used in the project for the development of other applications, namely laser Doppler vibrometry, gas sensing and a Datacom transceiver compatible with the 100GBase-LR4 standard. A key feature of this collaboration is the need for each partner to understand and to account for other elements in the supply chain. For example, device designers need to account for packaging when preparing chip layout, ensuring factors such as grating coupler pitch and shunt waveguides are included to ease the fibre alignment process. The PLAT4M project highlights the need for a consortium-driven approach to integrated photonics, where all elements of the technology supply chain work together to achieve a single objective.

ACKNOWLEDGMENTS

Funding is acknowledged by the European Union's Seventh Programme for research, technological development and demonstration under grant agreement No 318178.

AUTHOR CONTRIBUTIONS

M.A., J.B., C.L., E.L. and A.B. conceived and performed the experiments; T.S. and R.B. designed the Photonic Integrated Circuit; P.V. and P.A. fabricated the Photonic Integrated Circuit; C.S., J.S.L, C.E. and P.A.O.B. developed the device packaging; C.S., J.B., T.S. and P.A.O.B.wrote the paper.

CONFLICTS OF INTEREST

The authors declare no conflict of interest.

REFERENCES

1 Bruesselbach, H.; Jones, D.C.; Mangir, M.S.; Minden, M.I.; Rogers, J.L. Self-organized coherence in fibre laser arrays. *Opt. Lett.* **2005**, *30*, 1339–1341.

2 Corcoran, C.J.; Durville, F. Experimental demonstration of a phase-locked laser array using a self-Fourier cavity. *Appl. Phys. Lett.* **2005**, *86*, 1–3.

3 Shay, T.M.; Benham, V.; Baker, J.T.; Ward, C.; Sanchez, A.D.; Culpepper, M.A.; Pilkington, D.; Spring, J.; Nelson, D.J.; Lu, C.A. First experimental demonstration of self-synchronous phase locking of an optical array. *Opt. Express* **2006**,*14*, 12015–12021.

4 Augst, S.J.; Fan, T.Y.; Sanchez, A. Coherent beam combining and phase noise measurement of Yb fibre amplifiers. *Opt. Lett.* **2004**, *29*, 474–476.

5 Antier, M.; Bourderionnet, J.; Larat, C.; Lallier, E.; Lenormand, E.; Primot, J.; Brignon, A. kHz Closed Loop Interferometric Technique for Coherent Fibre Beam Combining. *IEEE J. Sel. Top. Quantum Electron.* **2014**, *20*, 182–187.

6 Vorontsov, M.; Weyrauch, T.; Lachinova, S.; Ryan, T.; Deck, A.; Gatz, M.; Paramonov, V.; Carhart, G. *Coherent Laser Beam Combining*; Wiley-VCH Verlag GmbH & Co. KGaA: Weinheim, Germany, 2013.

7 Vermeulen, D.; Selvaraja, S.; Verheyen, P.; Lepage, G.; Bogaerts, W.; Absil, P.; van Thourhout, D.; Roelkens, G. High-efficiency fibre-to-chip grating couplers realized using an advanced CMOS-compatible silicon-on-insulator platform. *Opt. Express* **2010**, *18*, 18278–18283.

8 Masood, A.; Pantouvaki, M.; Lepage, G.; Verheyen, P.; van Campenhout, J.; Absil, P.; van Thourhout, D.; Bogaerts, W. Comparison of heater architectures for thermal control of silicon photonic circuits. In Proceedings of the 2013 IEEE International Conference on Group IV Photonics (GFP), Seoul, Korea, 28–30 August 2013; pp. 83–84.

9 Spuesens, T.; Pathak, S.; Vanslembrouck, M.; Dumon, P.; Bogaerts, W. Grating Coupler with Integrated Grating Coupler/Power Splitter for On-chip Optical Power Distribution. In Proceedings of the IEEE 11th International Conference on Group IV Photonics (GFP), Paris, France, 27–29 August 2014; pp. 141–142.

10 Absil, P.; de Heyn, P.; Chen, H.; Srinivasan, A.; Verheyen, P.; Balakrishnan, S.; Lepage, G.; Pantouvaki, M.; de Coster, J.; Roelkens, G.; *et al.* Isipp-200: A silicon photonics platform supporting optical data rates beyond 50 Gb/s. *SPIE Photonics West* **2016**. in press.

11 Pavarelli, N.; Lee, J.S.; Rensing, M.; Scarcella, C.; Zhou, S.; Ossieur, P.; O'Brien, P.A. Optical and Electronic Packaging Processes for Silicon Photonic Systems. *J. Lightwave Technol.* **2015**, *33*, 991–997.

12 Taillaert, D.; van Laere, F.; Ayre, M.; Bogaertz, W.; van Thourhout, D.; Bienstman, P.; Baets, R. Grating Couplers for Coupling between Optical Fibres and Nanophotonic Waveguides. *Jpn. J. Appl. Phys.* **2006**, *45*, 6071–6077.

13 Bellanger, C.; Brignon, A.; Toulon, B.; Primot, J.; Bouamrane, F.; Bouvet, T.; Megtert, S.; Quétel, L.; Allain, T. Design of a fiber-collimated array for beam combining. *Opt. Eng.* **2011**, *50*.

CHAPTER 5

InP-Based Waveguide Triple Transit Region Photodiodes for Hybrid Integration with Passive Optical Silica Waveguides

Vitaly Rymanov *, Besher Khani, Sebastian Dülme, Peng Lu and Andreas Stöhr

Department of Optoelectronics/ZHO, University of Duisburg-Essen, Lotharstr. 55, Duisburg 47057, Germany

ABSTRACT

We report on a novel InP-based 1.55 μm waveguide triple transit region photodiode (TTR-PD) structure for hybrid integration with passive optical silica waveguides. Using the beam propagation method, numerical analyses reveal that, for evanescent optical coupling between a passive silica waveguide and the InP-based waveguide TTR-PD, a coupling efficiency of about 90% can be obtained. In addition to that, an absorption of about 50% is simulated within a TTR-PD length of 30 μm. For fabricated TTR-PD chips, a polarization dependent loss (PDL) of less than 0.9 dB is achieved within the complete optical C-band. At the operational wavelength of 1.55 μm, a reasonable PDL of 0.73 dB is measured. The DC responsivity and the RF responsivity are achieved on the order of 0.52 A/W and 0.24 A/W, respectively. Further, a 3 dB bandwidth of 132 GHz is experimentally demonstrated and high output-power levels exceeding 0 dBm are obtained. Even at the 3 dB cut-off frequency, no saturation effects occur at a photocurrent of 15.5 mA and an RF output power of −4.6 dBm is achieved. In addition to the numerical and experimental achievements, we propose a scheme for a hybrid-integrated InP/silicon-based photonic millimeter wave transmitter.

KEYWORDS

photonic integrated circuits; silicon photonics hybrid integration; heterogeneous integration; indium phosphide photonics; silica and silicon nitride photonics; optical interconnects; telecommunications; microwave photonics

1. INTRODUCTION

Research activities targeting millimeter wave (mmW) applications, *i.e.*, applications utilizing electromagnetic waves within the frequency range from 30 GHz to 300 GHz, have been revived in recent years. Many modern applications in sensing [1], imaging [2], radar [3], and, most notably, in wireless communications [4,5], make use of ever higher frequencies. One prominent example is that of future 5G mobile systems, which are currently being considered to make use of carrier frequencies in the mmW range [6,7]. This has simultaneously revived the interest in solid-state components and it has led to scores of high-performance components enabling, for example, highly-efficient mmW signal generation and detection [8]. In addition to purely electrical solutions using advanced InP- or SiGe-based solid-state amplifiers and mixers, integrated photonic technology has also become more mature, leading to photonic solutions ready for real-life implementations, e.g., the generation and distribution of low-phase-noise mmW local oscillator signals in radio astronomy [9]. Additionally, research on broadband wireless communications has led to recent outstanding results using photonic technologies. In particular, radio-over-fiber (RoF) techniques have enabled the transmission of extreme high-data (100 Gbit/s) wireless signals within mmW bands [10].

High-speed photodiodes (PDs) are key in such broadband RoF wireless networks. To meet the challenging demands for advanced mmW PDs in terms of bandwidth or output power, various groups are researching evolutionary, but also revolutionary, concepts helping to improve the performances of state-of-the-art mmW PDs [1,5,8,11,12,13,14,15,16,17,18]. Key objectives of this research are to expand the operational bandwidth of the PDs and simultaneously attain higher power-saturation levels [1]. High-frequency operation typically leads to PD structure designs including relatively thin absorptive layers to enhance the transit-time-limited response, which on the other hand simultaneously deteriorates the external quantum efficiency, primarily in PDs with top-illumination [13,14]. Furthermore, the resistance-capacitance (RC) time constant becomes the dominating limitation in terms of bandwidth for thin, undoped absorbing layers. The necessity to simultaneously achieve high external quantum efficiencies, high operational bandwidths for broadband (or full-band) operation, as well as high radio frequency (RF) output-power levels, favor the use of waveguide-based PDs [15,16]. This is due to their flexibility in the layer structure and topology design. As an example, mushroom-type waveguide PDs offer a solution for improving RC limitations, enabling high-frequency operation [17].

In this work, we report on InP-based 1.55 μm triple transit region PDs (TTR-PDs) [18], which employ the waveguide approach for hybrid integration with passive optical silica waveguides. As a future application (e.g., for 5G mobile systems), we first propose a scheme of a hybrid-integrated photonic mmW transmitter for photonic RF beam forming and beam steering. Since this approach is based on heterogeneous integration of InP- and silicon-based photonics components, the optical waveguide coupling is to be investigated for connecting the optical beam-steering network (including passive optical silica waveguides) [19,20] with a linear TTR-PD array. One of the challenges of these hybrid-integrated waveguide devices remains in the design of an appropriate layer structure, which enables efficient coupling of the optical mode from the passive silica waveguide into the active InP-based waveguide. To study evanescent coupling, we introduce an active waveguide based on the TTR-PD layer structure [18] on top of a passive optical silica waveguide [21]. Numerical simulations using the beam propagation method (BPM) are accomplished to demonstrate the optical TE-mode propagation in the hybrid-integrated InP/silica double waveguide. For experimental characterization, several TTR-PD chips were fabricated with monolithically-integrated passive optical waveguides to couple light into the active PD structure and, thus, measure the responsivity (DC and RF), the 3 dB bandwidth, and the 1 dB compression point of the TTR-PDs prior to further hybrid integration with silica waveguides. Accordingly, we report high-power (>0 dBm) TTR-PDs with a measured 3 dB bandwidth beyond 130 GHz.

2. SCHEME OF A HYBRID-INTEGRATED INP/SILICON-BASED PHOTONIC MILLIMETER WAVE TRANSMITTER

For future 5G mobile systems considered to make use of carrier frequencies in the mmW range, e.g., in the 60 GHz or the 70/80 GHz frequency band [4,6,7], photonic mmW transmitters that simultaneously enable high output-power levels, high operational bandwidth, and the opportunity for photonic beam forming, as well as beam steering, are necessitated. For this reason, a hybrid integration of different solid-state technologies on one platform becomes key for such high-performance devices. One of the benefits for a hybrid-integrated waveguide PD would be, for example, the low propagation loss (<0.1 dB/cm) [21] of the passive optical silica waveguides used in optical beam-steering networks [19,20]. A schematic view of a hybrid-integrated InP/silicon-based photonic mmW transmitter is illustrated in Figure 1.

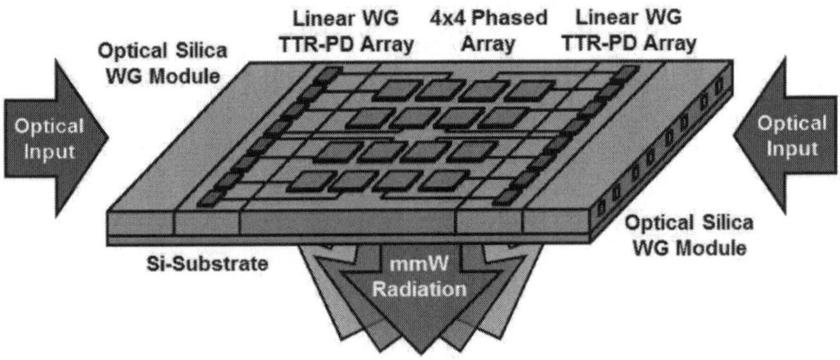

Figure 1. Schematic view and operation principle of a hybrid-integrated InP/silicon-based photonic millimeter wave (mmW) transmitter featuring two optical silica waveguide (WG) modules, two linear WG TTR-PD arrays and a 4 × 4 phased array antenna for mmW beam forming and beam steering.

As depicted in the scheme, the proposed photonic mmW transmitter is composed of different optics (optical waveguide modules), optoelectronics (waveguide TTR-PD arrays), and RF (phased array) sections. Two optical silica waveguide modules are integrated for double-sided illumination at the optical wavelength of 1.55 μm. An optical fiber array with eight conventional (*i.e.*, cleaved) single-mode fibers (SMFs) is considered for coupling of light into each silica module, which comprises monolithically-integrated spot-size converters [21] and an optical beam-steering network, e.g., using the true-time-delay approach [19,20].

The optical inputs of the linear TTR-PD array are interfaced with the corresponding optical outputs of the passive silica waveguides from the silica module. To couple the light vertically into the active TTR-PD structure, the TTR-PD is to be integrated on top of the passive silica waveguide. In the TTR-PD, the optical signal with the appropriate beam-steering signal (*i.e.*, phase) gets absorbed and converted into an electrical (mmW) signal that is further fed to an integrated antenna element, which is located on a silicon platform. Each TTR-PD allows individually addressing an antenna element of the phased array. Similar integration approaches on silicon submounts were already reported in previous works on hybrid-integrated InP/silicon-based photonic transmitter modules employing highly-resistive silicon spacers and hemispherical lenses [22,23]. Finally, the relative phases of the generated mmW signals can be adjusted in such a way that an efficient mmW beam of the entire phased array is generated and steered in a preferential direction [19,20].

3. SIMULATION OF EVANESCENT OPTICAL COUPLING BETWEEN THE TTR-PD AND A PASSIVE SILICA WAVEGUIDE

As the TTR-PD is a key element in the photonic transmitter scheme, which converts the optical signal from the optical to the RF domain, evanescent optical coupling between a silica (SiO_2) waveguide and the TTR-PD was studied using beam propagation method (BPM) analyses.

3.1. Modeling of the Hybrid-Integrated InP/Silica Double Waveguide

The cross section of the hybrid-integrated InP/silica double waveguide, which consists of an active InP-based TTR-PD and a passive silica waveguide section, is depicted schematically in Figure 2. In contrast to the monolithic approach that was reported in [18], the mushroom-type TTR-PD structure is flipped and hybrid-integrated on top of the silica waveguide. In addition to that, the thickness of the highly p-doped InP layer is reduced from 800 nm (see [18]) to 400 nm. Here, it should be noted that the free carrier absorption in the highly p-doped InP cladding layer gets higher than in the slightly n-doped InP cladding layer of the monolithically-integrated InGaAsP/InP waveguide (see Section 3.2 and Section 4).

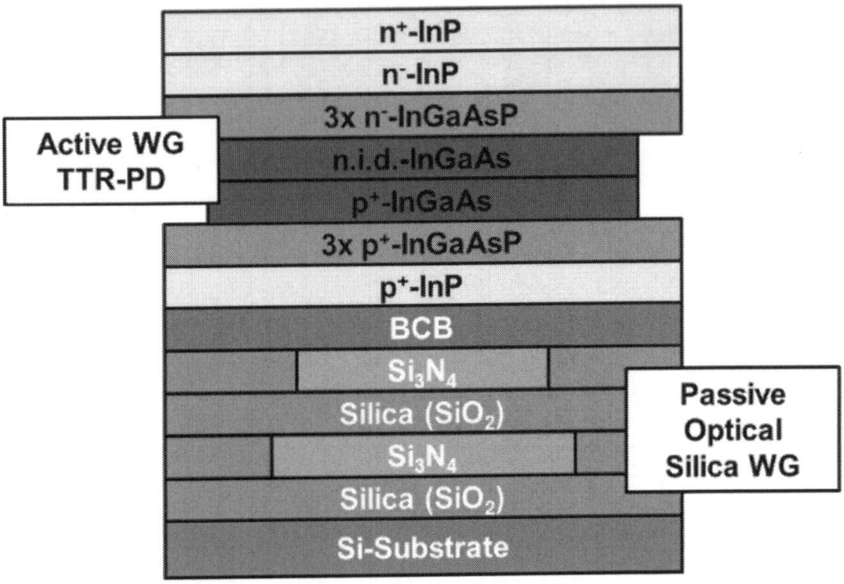

Figure 2. Schematic cross section of the developed InP/silica double waveguide comprising an active InP-based waveguide (WG) TTR-PD with a hybrid-integrated passive optical silica WG. Benzocyclobutene (BCB) is used for adhesive wafer bonding.

Further, the active TTR-PD structure comprises a 110 nm thick InGaAs core featuring a non-intentionally doped (n.i.d.) absorber layer (70 nm) and a highly p-doped absorber layer (40 nm) that are incorporated into the InGaAsP spacers and the InP cladding layers. The passive silica waveguide features two embedded silicon nitride (Si_3N_4) stripes and is grown on a Si-substrate. Details on both layer structures (*i.e.*, TTR-PD and silica waveguide) have been reported in [18,21]. To connect both waveguides, adhesive wafer bonding can be applied introducing a thin (<0.1 μm) intermediate benzocyclobutene (BCB) layer as BCB provides a refractive index of ~1.5 that is similar to the refractive index of silica (~1.45).

3.2. Numerical Beam Propagation Method Simulations

Based on the proposed layer structure from Figure 2, numerical BPM simulations for the propagation of TE-polarized light were accomplished for the developed InP/silica double waveguide (see Figure 3). To calculate the effective length of the TTR-PD with respect to coupling efficiency and absorption, the active coupling length was set to 100 µm.

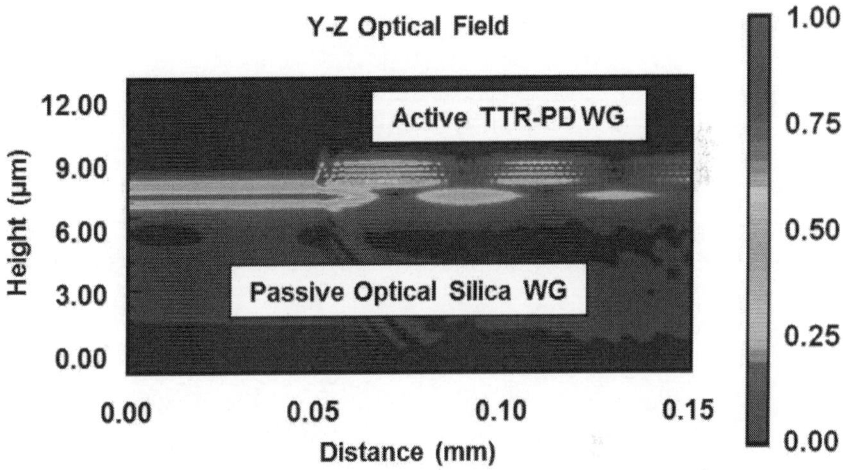

Figure 3. Simulated propagation of TE-polarized light demonstrating evanescent coupling between a silica waveguide (WG) and the InP-based TTR-PD WG (longitudinal cut).

From Figure 3, one can observe that a coupling efficiency of more than 90% within a coupling length of about 25 µm is achieved. It can further be observed that an absorption of about 50% is obtained in the InGaAs layers before the light couples into the silica waveguide (see Figure 3 and Figure 4). Additionally, the BPM simulations have indicated that around 14% of the propagation loss can be traced back to the free carrier absorption in the highly p-doped InP layer, which somewhat deteriorates the external quantum efficiency but actually does not affect the 3 dB bandwidth of the TTR-PD as the n.i.d.-InGaAs absorber layer, the slightly n-doped InGaAsP spacers, and the slightly n-doped InP collector of the TTR-PD remain unmodified, such as presented in [18].

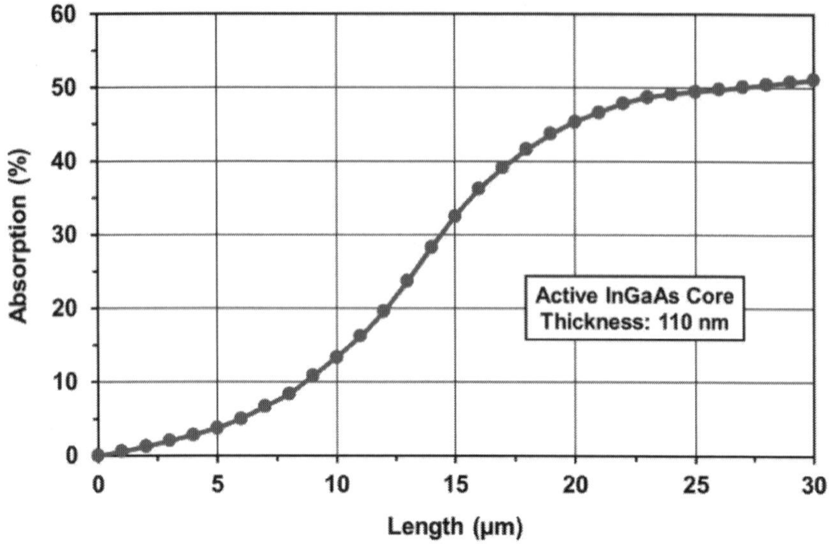

Figure 4. Simulated absorption at the optical wavelength of 1.55 μm plotted *versus* the length of the TTR-PD waveguide with an InGaAs core width of 2 μm.

It is worth mentioning that an absorption of around 80% could be achieved within the simulated coupling length of 100 μm, even though the absorption does not take place over the entire propagation term (see Figure 3). However, the extended length would raise the TTR-PD's junction capacitance, which is directly proportional to the length of the TTR-PD and, thus, deteriorates the 3 dB bandwidth.

4. EXPERIMENTAL RESULTS OF FABRICATED WAVEGUIDE TRIPLE TRANSIT REGION PHOTODIODE CHIPS

Prior to interfacing the TTR-PDs with integrated silica waveguides, the DC responsivity, the RF responsivity, the 3 dB bandwidth, and the 1 dB compression point of the developed TTR-PD structure were to be investigated. Based on the achieved BPM simulation results, several devices with an active TTR-PD area of 2 × 30 μm^2 were fabricated and experimentally characterized. To characterize the TTR-PDs, a monolithically-integrated passive optical InGaAsP/InP-based waveguide, providing a similar optical TE-mode profile with a mode field diameter (MFD) on the order of 1.5 μm [21], was substituted for the silica waveguide in the layer structure design to enable (evanescent) optical coupling to the TTR-PD chip (see Figure 5a). The top view of a fabricated TTR-PD chip is further shown in Figure 5b. For on-chip measurements, no additional cooling arrangements, such as Peltier elements or heat sink modules, were applied to the TTR-PD chips. Further, it should be noted that no anti-reflection coating was deposited at the passive optical waveguide facets.

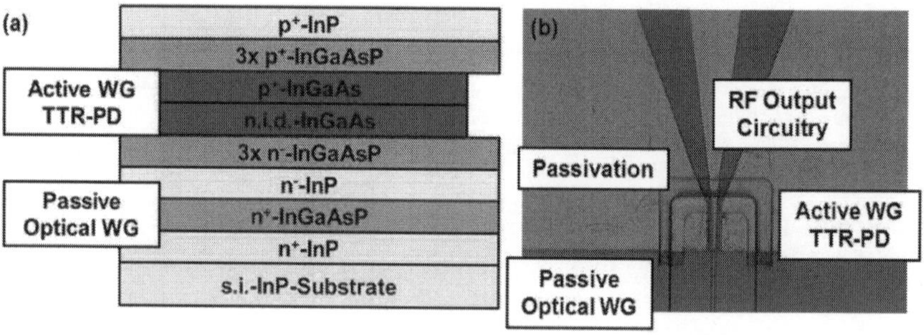

Figure 5. (**a**) Schematic cross section of the TTR-PD layer structure with a monolithically-integrated passive optical waveguide (WG) and (**b**) scanning electron microscope (SEM) photograph of the fabricated WG TTR-PD chip.

4.1. Polarization Behavior and DC Responsivity

To investigate the polarization behavior for optical heterodyne mmW signal generation (see Section 4.2), the polarization dependent loss (PDL) of the TTR-PDs was measured within the complete optical C-band from 1530 to 1565 nm (see Figure 6). The optical fiber-chip coupling to the passive optical waveguide was accomplished using a lensed SMF. The applied SMF featured a spot size diameter on the order of 1.7 μm to reduce the optical mode mismatch, which would be induced by using a standard SMF with an MFD of about 10.4 μm. For measuring the PDL, the optical input-power level was set to 10 dBm. Two direct current (DC) needle manipulators were used for biasing the TTR-PD chip at a bias voltage of −8 V. The measured PDL is plotted in Figure 6.

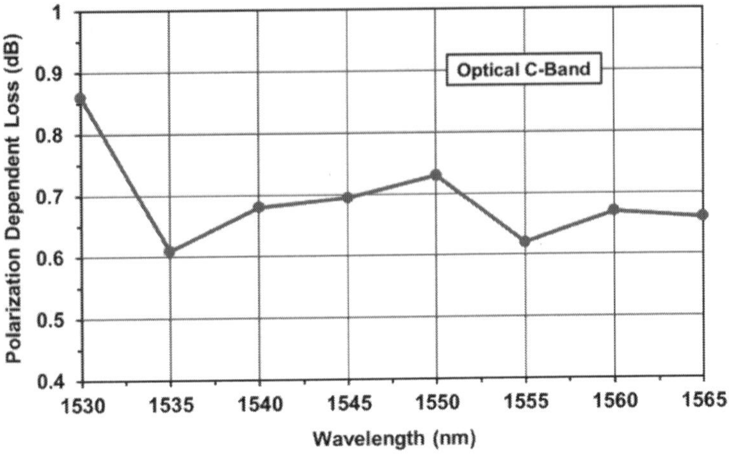

Figure 6. Measured polarization dependent loss plotted against the optical wavelength for a bias voltage of −8 V.

As can be seen from Figure 6, within the entire C-band the TTR-PDs exhibit a reasonable PDL, which is less than 0.9 dB. A PDL of 0.73 dB can further be observed at the operational wavelength of 1.55 µm. In addition to that, a peak DC responsivity ($R_{DC,max}$) on the order of 0.52 A/W was achieved (see Figure 7).

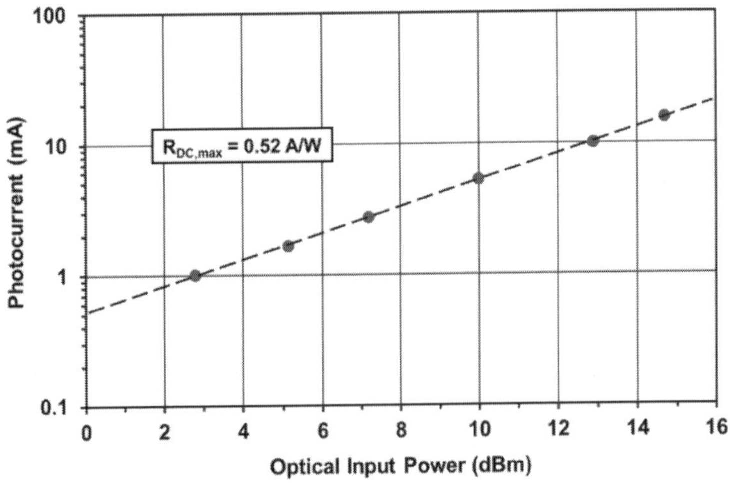

Figure 7. Measured photocurrent *versus* the optical input power for the operational wavelength of 1.55 µm at a bias voltage of −8 V. The dashed line shows the simulated photocurrent for the DC responsivity of 0.52 A/W.

4.2. Measured Frequency Response

For measuring a frequency response from DC up to 220 GHz, we used a calibrated scalar network analyzer setup featuring a tunable external cavity laser (TLS) in conjunction with a polarization controller (PC) and a continuous wave laser source (LS) for optical heterodyne mmW signal generation. The measurement setup is illustrated in Figure 8. Three different 100 μm pitch coplanar RF probes (~50 Ω) were used to attain on-chip power measurements within the measured frequency range shown in Figure 9.

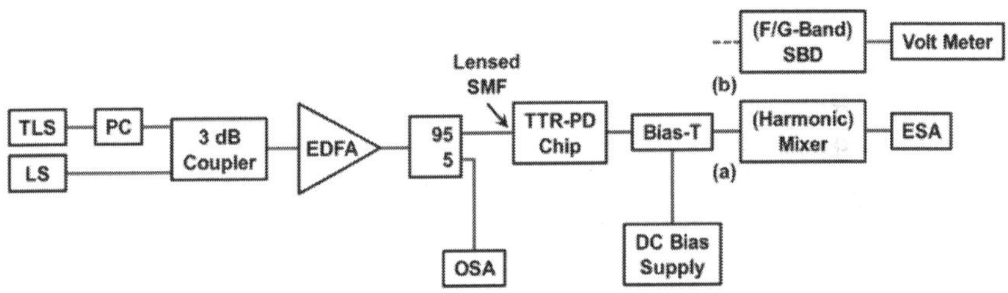

Figure 8. Measurement setup based upon optical heterodyne millimeter wave signal generation up to 220 GHz featuring (**a**) an electrical spectrum analyzer (ESA) with internal and external mixers (DC–110 GHz) and (**b**) Schottky barrier diodes (SBDs) connected to a volt meter (90–220 GHz). An erbium-doped fiber amplifier (EDFA) and an optical spectrum analyzer (OSA) are further included in the setup.

Up to a frequency of 110 GHz, an RF probe and an external bias-T, both including a 1 mm coaxial W1-connector, were applied together with an electrical spectrum analyzer (ESA) employing an internal mixer (up to 50 GHz) and the external V-band (50–75 GHz) and W-band (75–110 GHz) harmonic mixers. Within the frequency ranges of 90–140 GHz (F-band) and 140–220 GHz (G-band), two further RF probes featuring a WR-08 and a WR-05 waveguide output, respectively, and an integrated bias-T were used with corresponding Schottky barrier diodes (SBDs) and a volt meter. The measured frequency response up to 190 GHz is presented in Figure 9. The photocurrent and the bias voltage of the TTR-PD were set to 1 mA and −8 V, respectively. From Figure 9, one can notice that a 3 dB cut-off frequency of 132 GHz is achieved. The resulting RF responsivity is found to be on the order of 0.14 A/W. At lower frequencies, a maximum RF responsivity of about 0.24 A/W has been achieved.

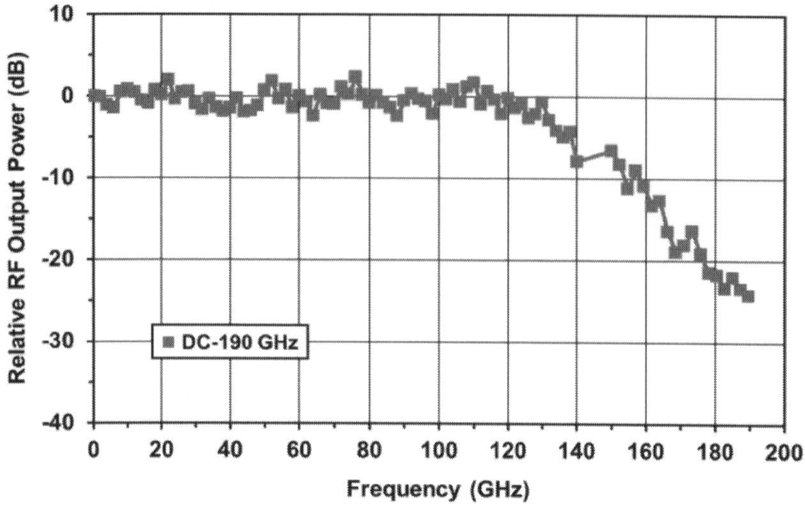

Figure 9. Measured relative RF output power from DC up to 190 GHz.

4.3. RF Output Power Measurements

Furthermore, the saturation effects were investigated within the 3 dB bandwidth. For calibrated power measurements at the 3 dB cut-off frequency (132 GHz), the F-band (90–140 GHz) SBD including an integrated WR-08 waveguide has been used together with a level set attenuator as the saturation of the SBDs starts at an RF input power of about −20 dBm. The maximum attenuation of the attenuator was set to 20 dB. Thus, the detected RF output power was measured for different optical input-power levels. In Figure 10, the measured RF output power is plotted against the photocurrent for the bias voltage of −8 V.

As shown in Figure 10, even at the 3 dB cut-off frequency no saturation effects occur and a maximum RF output power of −4.6 dBm is achieved, along with a photocurrent level of 15.5 mA. To identify the 1 dB compression point of the TTR-PD, a higher optical input-power level, which was not obtainable by the measurement setup, was required. However, the catastrophic optical damage of the passive optical waveguide facet caused by the applied lensed SMF has been calculated for an optical input-power level of about 24 dBm [24]. Maximum RF output-power levels exceeding 0 dBm (e.g., at 110 GHz) were attained as can be seen from Figure 10.

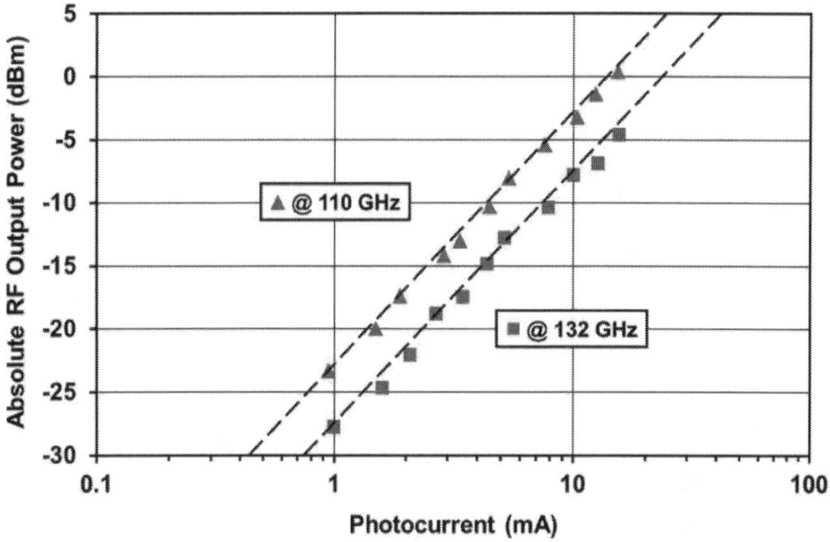

Figure 10. Measured RF output power plotted against the photocurrent for a bias voltage of −8 V at the frequencies of 110 GHz and 132 GHz. The dashed lines show the simulated output power for the RF responsivities of 0.24 A/W (110 GHz) and 0.14 A/W (132 GHz).

For hybrid integration of TTR-PDs on top of an optical silica waveguide, a limited number of modified TTR-PD chips have already been fabricated. Details will be reported elsewhere.

5. CONCLUSIONS

In this paper, we presented a novel high-speed InP-based waveguide triple transit region photodiode (TTR-PD) structure for hybrid integration with passive optical silica waveguides, which are fundamental to form an optical beam-steering network. As an application for future 5G mobile systems, we proposed a scheme for a hybrid-integrated InP/silicon-based photonic millimeter wave transmitter, enabling photonic RF beam forming and beam steering. For this approach, the TTR-PD chip was integrated on top of a passive optical silica waveguide to achieve evanescent optical coupling. Numerical beam propagation method analyses revealed that a coupling efficiency of around 90% is attained between a passive silica waveguide and the active InP-based waveguide TTR-PD, as well as an absorption of about 50% within a simulated TTR-PD length of 30 μm. For fabricated TTR-PD chips including a monolithically-integrated passive optical waveguide, a reasonable polarization dependent loss of less than 0.9 dB was achieved within the optical C-band (e.g., 0.73 dB at 1.55 μm). Moreover, a 3 dB cut-off frequency of 132 GHz was experimentally demonstrated. Within the 3 dB bandwidth, RF output-power levels higher than 0 dBm were measured for the fabricated TTR-PDs. The DC responsivity and the RF responsivity were found to be on the order of 0.52 A/W and 0.24 A/W, respectively.

ACKNOWLEDGMENTS

This work was supported by the European STReP IPHOBAC-NG (www.iphobac-ng.eu) under the grant No. 619870.

AUTHOR CONTRIBUTIONS

V.R., S.D., and A.S. developed and designed the component. V.R., P.L., and S.D. performed the simulations. V.R. developed the fabrication process and fabricated the devices. V.R., S.D., and B.K. performed the experimental characterization. V.R., B.K., and A.S. wrote the paper, supported by all other coauthors.

CONFLICTS OF INTEREST

The authors declare no conflict of interest.

REFERENCES

1 Nagatsuma, T.; Ito, H.; Ishibashi, T. High-power RF photodiodes and their applications. *Laser Photonics Rev.* **2009**, *3*, 123–137.

2 Ahmed, S.S.; Schiessl, A.; Schmidt, L. A novel fully electronic active real-time imager based on a planar multistatic sparse array. *IEEE Trans. Microw. Theory Tech.* **2011**, *59*, 3567–3576.

3 Tseng, T.-F.; Wun, J.-M.; Chen, W.; Peng, S.-W.; Shi, J.-W.; Sun, C.-K. High-depth-resolution 3-dimensional radar-imaging system based on a few-cycle W-band photonic millimeter-wave pulse generator. *Opt. Express* **2013**, *21*, 14109–14119.

4 Stöhr, A. Photonic millimeter-wave generation and its applications in high data rate wireless access. In Proceedings of the IEEE International Topical Meeting on Microwave Photonics, Montreal, QC, Canada, 5–9 October 2010; pp. 7–10.

5 Seeds, A.J.; Shams, H.; Fice, M.J.; Renaud, C.C. Terahertz photonics for wireless communications. *J. Lightwave Technol.* **2015**, *33*, 579–587.

6 METIS. Available online: https://www.metis2020.com/ (accessed on 19 November 2015).

7 RAPID5G. Available online: http://www.rapid-5g.eu/ (accessed on 19 November 2015).

8 Stöhr, A.; Babiel, S.; Cannard, P.J.; Charbonnier, B.; van Dijk, F.; Fedderwitz, S.; Moodie, D.; Pavlovic, L.; Ponnampalam, L.; Renaud, C.C.; *et al.* Millimeter-wave photonic components for broadband wireless systems. *IEEE Trans. Microw. Theory Tech.* **2010**, *58*, 3071–3082.

9 ALMA. Available online: http://www.almaobservatory.org/ (accessed on 19 November 2015).

10 König, S.; Lopez-Diaz, D.; Antes, J.; Boes, F.; Henneberger, R.; Leuther, A.; Tessmann, A.; Schmogrow, R.; Hillerkuss, D.; Palmer, R.; *et al.* Wireless sub-THz communication system with high data rate. *Nat. Photonics* **2013**, *7*, 977–981.

11 Khani, B.; Rymanov, V.; Steeg, M.; Buck, A.; Dülme, S.; Stöhr, A. Compact E-band (71–86 GHz) bias-tee module for external biasing of millimeter wave photodiodes. In Proceedings of the IEEE International Topical Meeting on Microwave Photonics, Paphos, Cyprus, 26–29 October 2015.

12 Wun, J.-M.; Liu, H.-Y.; Lai, C.-H.; Chen, Y.-S.; Yang, S.-D.; Pan, C.-L.; Bowers, J.E.; Huang, C.-B.; Shi, J.-W. Photonic high-power 160-GHz signal generation by using ultrafast photodiode and a high-repetition-rate femtosecond optical pulse train generator. *IEEE J. Sel. Top. Quantum Electron.* **2014**, *20*, 10–16.

13 Urick, V.J.; McKinney, J.D.; Williams, K.J. High current photodetectors. In *Fundamentals of Microwave Photonics*, 1st ed.; John Wiley & Sons, Inc.: Hoboken, NJ, USA, 2015; pp. 351–382.

14 Preu, S.; Döhler, G.H.; Malzer, S.; Stöhr, A.; Rymanov, V.; Göbel, T.; Brown, E.R.; Feiginov, M.; Gonzalo, R.; Beruete, M.; *et al.* Principles of THz generation. In *Semiconductor Terahertz Technology: Devices and Systems at Room Temperature Operation*,

1st ed.; Carpintero, G., García Muñoz, L.E., Hartnagel, H.L., Preu, S., Räisänen, A.V., Eds.; John Wiley & Sons, Ltd.: Chichester, UK, 2015; pp. 3–68.

15 Rouvalis, E.; Chtioui, M.; Tran, M.; Lelarge, F.; van Dijk, F.; Fice, M.J.; Renaud, C.C.; Carpintero, G.; Seeds, A.J. High-speed photodiodes for InP-based photonic integrated circuits. *Opt. Express* **2012**, *20*, 9172–9177.

16 Beling, A.; Cross, A.S.; Piels, M.; Peters, J.; Fu, Y.; Zhou, Q.; Bowers, J.E.; Campbell, J.C. High-power high-speed waveguide photodiodes and photodiode arrays heterogeneously integrated on silicon-on-insulator. In Proceedings of the Optical Fiber Communication Conference and Exposition and the National Fiber Optic Engineers Conference, Anaheim, CA, USA, 17–21 March 2013; pp. 1–3.

17 El-Batawy, Y.M.; Youssef, S. Biasing effects on the characteristics and the optimization of mushroom waveguide photodetectors (WGPDs). In Proceedings of the International Symposium on Photonics and Optoelectronics (SPIE 9233), Suzhou, China, 22–24 August 2014.

18 Rymanov, V.; Stöhr, A.; Dülme, S.; Tekin, T. Triple transit region photodiodes (TTR-PDs) providing high millimeter wave output power. *Opt. Express* **2014**, *22*, 7550–7558.

19 Meijerink, A.; Roeloffzen, C.G.H.; Meijerink, R.; Zhuang, L.; Marpaung, D.A.I.; Bentum, M.J.; Burla, M.; Verpoorte, J.; Jorna, P.; Hulzinga, A.; *et al.* Novel ring resonator-based integrated photonic beamformer for broadband phased array receive antennas—Part I: Design and performance analysis. *J. Lightwave Technol.* **2010**, *28*, 3–18.

20 Zhuang, L.; Roeloffzen, C.G.H.; Meijerink, A.; Burla, M.; Marpaung, D.A.I.; Leinse, A.; Hoekman, M.; Heideman, R.G.; van Etten, W. Novel ring resonator-based integrated photonic beamformer for broadband phased array receive antennas—Part II: Experimental prototype. *J. Lightwave Technol.* **2010**, *28*, 19–31.

21 Wörhoff, K.; Heideman, R.G.; Leinse, A.; Hoekman, M. TriPleX: A versatile dielectric photonic platform. *Adv. Opt. Technol.* **2015**, *4*, 189–207.

22 Babiel, S.; Rymanov, V.; Stöhr, A. Hybrid integrated photonic millimeter wave emitter. In Proceedings of the Asia-Pacific Microwave Photonics Conference, Kyoto, Japan, 25–27 April 2012.

23 Rymanov, V.; Dülme, S.; Babiel, S.; Stöhr, A. Compact photonic millimeter wave (200–300 GHz) transmitters based on semicircular bow-tie antenna-integrated 1.55 μm triple transit region photodiodes. In Proceedings of the German Microwave Conference, Aachen, Germany, 10–12 March 2014; pp. 1–4.

24 Rymanov, V.; Khani, B.; Dülme, S.; Steeg, M.; Stöhr, A. Compact triple transit region photodiode module with WR-12 rectangular waveguide output. In Proceedings of the IEEE International Topical Meeting on Microwave Photonics, Paphos, Cyprus, 26–29 October 2015.

CHAPTER 6

Quantum Dot/Liquid Crystal Nanocomposites in Photonic Devices

Andrea L. Rodarte [1], Fredy Cisneros [1], Jason E. Hein [2,*], Sayantani Ghosh [1,*] and Linda S. Hirst [1,*]

[1] Department of Physics, University of California, 5200 Lake Rd, Merced, CA 95343, USA

[2] Department of Chemistry, University of British Columbia, 2036 Main Mall, Vancouver, BC V6T1Z1, Canada

ABSTRACT

Quantum dot/liquid crystal nano-composites are promising new materials for a variety of applications in energy harvesting, displays and photonics including the liquid crystal laser. To realize many applications, however, we need to control and stabilize nano-particle dispersion in different liquid crystal host phases and understand how the particles behave in an anisotropic fluid. An ideal system will allow for the controlled assembly of either well-defined nano-particle clusters or a uniform particle distribution. In this paper, we investigate mesogen-functionalized quantum dots for dispersion in cholesteric liquid crystal. These nanoparticles are known to assemble into dense stable packings in the nematic phase, and such structures, when localized in the liquid crystal defects, can potentially enhance the coupling between particles and a cholesteric cavity. Controlling the dispersion and assembly of quantum dots using mesogenic surface ligands, we demonstrate how resonant fluid photonic cavities can result from the co-assembly of luminescent nanoparticles in the presence of cholesteric liquid crystalline ordering.

KEYWORDS

quantum dot; liquid crystal; cholesteric

1. INTRODUCTION

One of the most interesting problems in nano-science in recent years has been to find a way to organize nanoparticles of different types into defined distributions. This could be close packed assemblies, dynamic clusters of different particles or well dispersed distributions of particles in a fluid phase. The goal is to produce a macroscopic composite material with properties that emerge from the particular controlled nanoparticle arrangement within. Such meta-materials are predicated to exhibit effects such as low or negative refractive indices, superparamagnetism, metal enhanced fluorescence and localized Plasmon resonance [1,2]. Nanoparticles can be difficult to disperse and control in an assembly due to the strong Van der Waals attractions between nano-scale particles. Uncontrolled assembly is a significant problem when surface interactions come to dominate at the nanoscale and surface chemistry strategies are usually employed to moderate these strong short-range surface forces. A common strategy is to use alkyl chain ligands such as octadecylamine (ODA). These chains provide a steric shield against irreversible aggregation by taking advantage of the high entropic cost to chain interdigitation.

Recently, advances have been made in the controlled assembly of very well ordered 2D nanoparticle lattices on solid substrates [3]. These assemblies are assisted by ligand-ligand interactions between the particles. There are many unique applications for nanoparticle lattices, but the production of these materials on a large scale is a challenge. In addition, the particles need to be monodispersed. In a different approach, cluster assembled materials (3D disordered clusters of particles) can provide many of the advantages of ordered assemblies such as cooperative effects but with a more scalable approach and less strict dispersion requirements. Solution based assembly methods for particle clusters can be generalized to include many different particle combinations and recent attempts at assemblies of gold particle clusters [4] have used a variety of particle spacing methods including biopolymers [5,6] proteins [7,8] and block copolymers [9,10,11].

In this paper, we report a comparison study between two different Quantum dot (QD)/cholesteric liquid crystal (LC) composites. We will discuss our approaches to producing liquid crystal-based materials with defined nanoparticle distributions, focusing specifically on quantum dots owing to their unique size-tunable luminescent properties. Nanoparticle ligands can be designed to respond to the ordered molecular environment of a liquid crystal phase, inducing dispersion [12], clustering [13] or assembly at the interface of a liquid crystal phase transition [14]. These techniques can provide easily scalable methods for producing a rich variety of quantum dot and nanoparticles assemblies. In particular, we present spectral characterization of quantum dots embedded in a cholesteric liquid crystal cavity with two different surface ligands.

Liquid crystals are ordered fluids characterized by molecular anisotropy. The most well known example of a liquid crystal phase is the nematic phase, typically formed from rod-like molecules in which the constituent molecules have local orientational order, defined by a director, c. The nematic phase is widely used in the display industry, for which molecular reorientation under an applied electric field provides the mechanism for optical switching. There are many possible liquid crystal phases, with varying degrees of positional and orientational order, but still maintaining fluid-like properties and these include the layered smectic phases, columnar phases with 2D hexagonal order and complex chiral phases such as the blue phases. The cholesteric phase, a chiral analog of the nematic phase is particularly interesting for photonic devices as it exhibits a 1D-photonic band gap in planar alignment that may be useful for lasing applications [15] and spontaneous emission enhancement in single-photon sources for secure quantum communication [16,17]. Although much more complicated, the blue phases (a 3D lattice of twisted cylinders) also exhibit photonic band gaps and have been explored for similar lasing applications.

One of the strengths of using liquid crystal materials in a photonic device is their structural tunability. The photonic band gap of the cholesteric phase is tunable with temperature and applied electric field. In addition, the phase can be designed to have a specific band gap over a desired temperature range by varying its molecular composition. Since liquid crystals can be designed to exhibit phases of interest at and around room temperature, material mixing and processing is relatively easy. Liquid crystals can be used to create electro-optically active thin films and apply them to a variety of different device geometries without the constraints of solid material growth or lithography. This ordered fluid medium provides an ideal environment to create a dispersion of QDs for printable photonic devices. Once assembled, the materials containing nanoparticle structures or dispersions can be applied to a substrate and polymerized if desired to produce a permanent film or coating. Applications for QDs dispersed in fluid media include solar concentrators and printable laser inks.

2. EXPERIMENTAL SECTION

In this section, we first describe our methods for preparing composite cholesteric/QD materials, then, in the following results section, we demonstrate how resonant fluid photonic cavities can result from the co-assembly of luminescent nanoparticles in the presence of cholesteric liquid crystalline ordering.

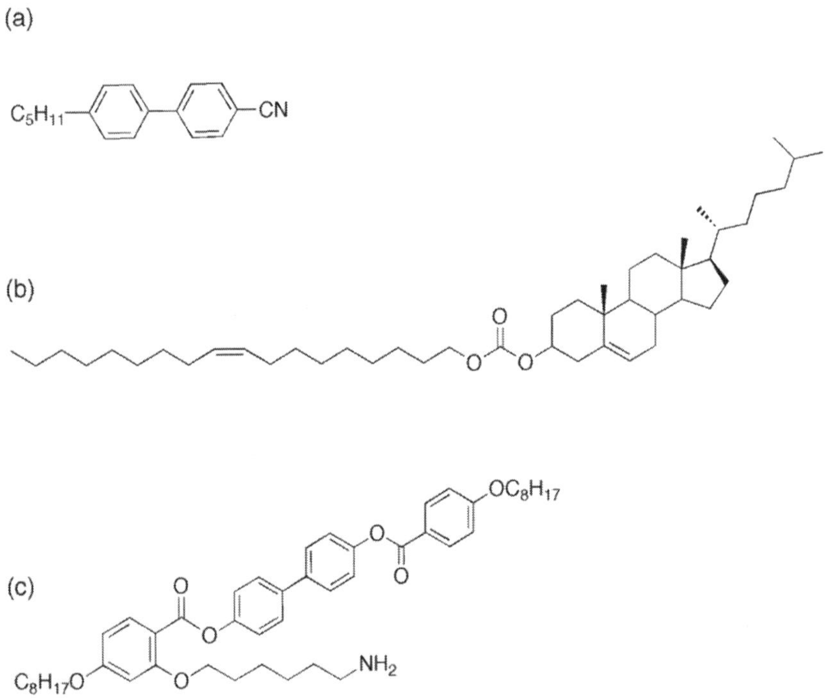

Figure 1. Liquid crystal compounds used in this study (**a**) 4-Cyano-4'-pentylbiphenyl (5CB), (**b**) Cholesterol oleyl carbonate (COC) and (**c**) the mesogenic ligand (L1) [18].

We start with commercially available quantum dots with octadecylamine (ODA) surface ligands (5.2–6.2 nm, NN Labs Inc) (ODA-QDs) onto which mesogenic ligands can be exchanged. Synthesis of the mesogenic ligand used in this paper was carried out as previously reported [18], Figure 1c shows the molecular structure of the ligand. These ligands exhibit liquid crystalline phases in their pure state with high clearing points above 100 °C—a contrast with the host phase (5CB (4-cyano-4'-pentylbiphenyl)), which exhibits the nematic to isotropic transition at 34 °C (Figure 1a). As previously described [18], ligands are exchanged on to the surface of the CdSe/ZnS quantum dots to form LC–QDs. These new particles are then dispersed into the cholesteric liquid crystal by wt%. Bath sonication for eight hours at 40 °C in the isotropic phase ensures an excellent initial dispersion, verified by fluorescence microscopy.

Cholesteric liquid crystal is prepared by mixing cholesterol oleyl carbonate (COC, Sigma Aldrich, Figure 1b and the nematic liquid crystal, 4'-pentyl'4-biphenylcarbonitrile (5CB, Sigma Aldrich) in varying proportions by weight. The mixtures were sonicated at 43 °C (above the isotropic transition temperature for both compounds) for 1 h to ensure good mixing. CdSe/ZnS QDs with octadecylamine ligands and an emission in toluene solution centered at 609 nm, were added to the LC mixtures at a concentration of 0.01–0.02 wt%, then sonicated again for 1–6 h to obtain the final QD–LC composite.

Fluorescence microscopy is used to image the spatial distribution of quantum dots in the host phase. In all experiments presented here we used CdSe/ZnS core shell quantum dots with a 620 nm emission peak. Fluorescence microscopy was carried out on a Leica DM2500P microscope in reflection mode using a 20× or 40× objective. For the 620 nm QDs a 510–560 band pass filter was used with a white light mercury lamp illumination. Emission was detected using a 580 nm dichroic mirror and a 590 nm long pass filter. The microscope can also be used in transmission mode with the sample between crossed polarizers to image liquid crystal birefringence. Samples were mounted on standard glass microscope slides under a cover slip.

To prepare a planar alignment coating, glass slides were spin coated with a 1 wt% aqueous polyvinyl alcohol (PVA) solution (Sigma Aldrich), dried in air and rubbed to establish the required orientation. To prepare a homeotropic alignment layer, glass slides were dip coated for 5 min in 0.5 mM aqueous hexadecyltrimethylammonium bromide (CTAB, Sigma Aldrich) solution.

For QD PL measurements, a VERDI 532 nm continuous wave excitation laser is focused on the sample through a 100× Nikon objective to produce a diffraction-limited spot diameter of approximately 600 nm. Samples are mounted on a motorized high-resolution scanning stage, and spectra recorded with an Acton 300i Spectrometer that disperses the signal onto a thermo-electrically cooled CCD camera with a 0.18 nm resolution.

Samples were mounted on glass slides for characterization prepared with a planar alignment layer and initially heated to 50 °C into the isotropic phase. The slides were then cooled at a rate of 1 °C/min from the isotropic phase into the nematic phase and left to anneal at 30 °C overnight.

When the sample is excited with 532 nm laser light we analyze device emission via two collection fibers as shown inFigure 2 in order to probe the two circularly polarized components of the QD emission from the sample simultaneously. The circularly polarized light is first converted to a linearly polarized beam using a quarter wave plate, and then passed through a Glan laser polarizer that allows us to measure the intensities of both the s and the p polarizations. It is clear from the spectral data shown in the Figure that one component contains emission (pink) that is not affected by the reflection band (green), *i.e.*, the incident polarization was of opposite handedness to that of the cholesteric phase, and one that matches the cholesteric twist, resulting in emission enhancement effects. In the blue curve, we observe a small amount of coupling and amplification from the edge of the cavity, demonstrating cavity coupling with the embedded QDs.

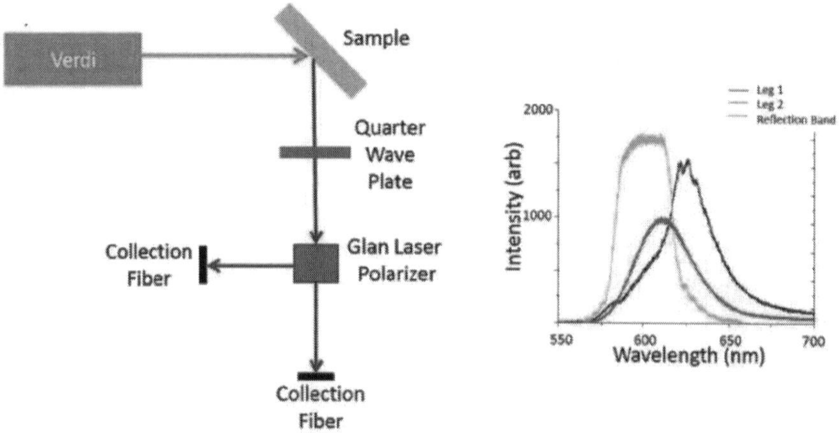

Figure 2. Schematic of setup used to collect polarization resolved QD emission, and a typical emission example of an emission spectrum from embedded QDs in the cholesteric phase showing both linear components of the emission overlaid on the reflection spectrum for the stop band of the cholesteric.

Controlling the dispersion of quantum dots, or other similarly sized nanoparticles in liquid crystal, thus forming a stable suspension without aggregation is an important goal for photonic applications. A stable distribution of particles may be desired but the materials can age—particles falling out of suspension and degrading device performance. Isotropic ligands such as ODA inherently produce defects in an oriented liquid crystal environment. A particle with uniform planar or homeotropic anchoring will induce a topological defect in the liquid crystal director field due to the frustration between local molecular alignment and anchoring conditions at the particle surface. We observe splay bend deformation of the local liquid crystal director as a result of nanoparticle inclusion in the nematic phase [13]. The free energy cost of these deformations is described by the Frank elastic constants. QDs and other nanoparticles tend to cluster together in the nematic phase to reduce this energy cost and control over this clustering effect is important to produce uniform stable materials. However, the formation of small nanoparticle clusters may offer some advantages for specific photonics applications.

3. RESULTS AND DISCUSSION

Figure 3a,c shows polarized optical microscopy images of the cholesteric texture for both the ODA–QD and LC–QDs embedded in the cholesteric phase along with fluorescence microscopy images of the same area showing QD distribution. One of the most apparent differences here seen in the optical microscopy is that at this concentration (0.1 wt%), the ODA–QD mixture exhibits much larger aggregates in the cholesteric liquid crystal (CLC) whereas the LC–QDs produce small but also distinct aggregates. This result is consistent with prior results [13,14] in which the nematic phase transition acts to assemble the included nanoparticles into templated structures at the phase transition. At 0.1 wt% and above, we expect aggregate formation for both of these particle types [18].

For particles with the LC ligand, large assemblies were recently reported to form when dispersed in the isotropic phase and cooled through the nematic phase transition. The mesogenic ligands may act to stabilize these structures (Figure 4) providing a mechanism to create hierarchical composite materials. Recently, our group developed a new methodology to assemble large-scale QD structures using the particles' spatial distribution across the nematic to isotopic phase transition [13,14]. Figure 4a,b shows a fluorescence image of QD "vesicles" or micro-shells stabilized by this process in which the particles are densely packed with an average separation of ~11 nm [14]. Quantum dot membranes an also be prepared by a similar process (Figure 4c) in which the nanoparticles are stabilized in an interfacially templated assembly.

When the devices shown in Figure 3 were excited with the 532 nm continuous wave laser, we observed characteristic emission for QDs coupling with the cholesteric cavity for both types of particle (Figure 3e,f). The results are consistent with recent reported data from our group for ODA–QDs only [19,20]. In the ODA–QD device we observed that good cavity coupling as shown in Figure 3e as an example was very location dependent. This should be expected for a material in which almost all of the emitting particles are located within dense aggregates using an excitation spot size of 600 nm. In comparison, the LC–QD device showed a much more uniform spatial performance. This result supports the idea that although aggregates can be seen optically, there are also many QDs dispersed throughout the phase.

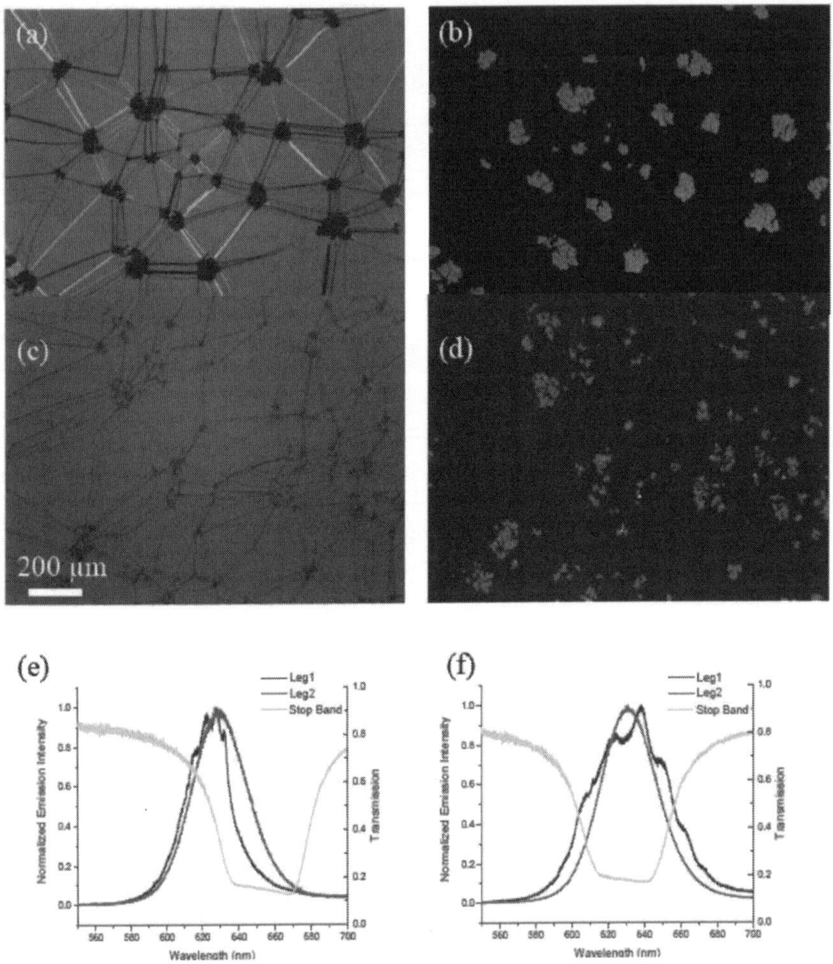

Figure 3. Polarized optical microscope images of 0.1 wt% mixtures of (**a**) ODA–QDs and (**c**) functionalized LC–QDs dispersed into a planar aligned cholesteric film shown next to corresponding fluorescence microscopy images (**b**) and (**d**). (**e**) and (**f**) show corresponding emission spectra for two different circular polarizations for the QDs embedded in these fluid cavities. Superimposed are transmission curves for circularly polarized light demonstrating the position of the stop band in each material.

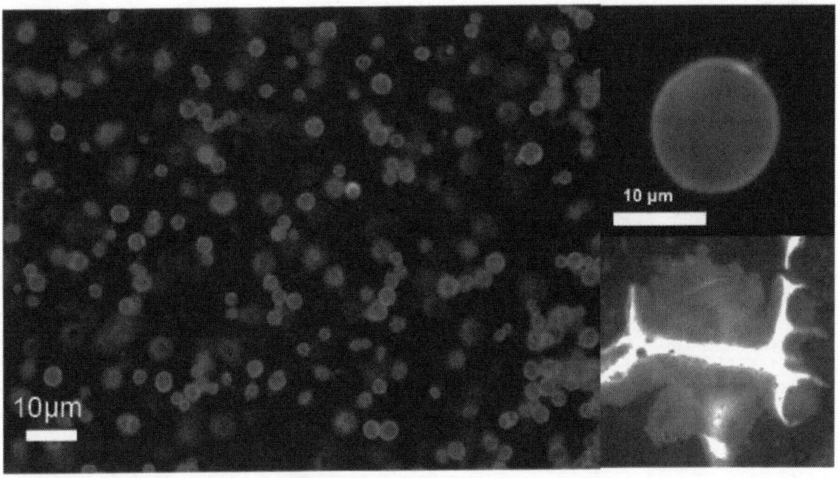

Figure 4. Fluorescence microscope images of quantum dot micro-shells, or "vesicles" and a flat membrane generated by phase transition templating (ref). (**a**) multiple shells suspended in the nematic phase, (**b**) a single QD shell and (**c**) a crumpled membrane structure formed by the same method.

In a second experiment, we investigated the effect of localizing the QDs outside of the cholesteric fluid when compared with embedding the particles. These comparison samples were prepared from films of ODA-QDs suspended in PMMA and toluene at 0.8 wt% QD in PMMA. The QD/PMMA solution was deposited on glass using a spin coater and cured at 140 °C for one hour. After curing, an alignment layer of PVA was deposited on top of the film as is done for clean glass. The film was then covered with 57.5%, 42.5% COC/5CB mixture to create a photonic cavity that overlaps with the QD emission.

We expected that the cavity coupling would be stronger for similar number densities of QDs emitting from within the cholesteric cavity when compared with emitters located outside the cavity. To test our hypothesis, we prepared two different 0.02 wt% mixtures of QDs in CLC, ODA–QDs and LC–QDs, functionalized with the mesogenic ligand shown inFigure 1c. Such a low concentration of QDs was used in this case to ensure a uniform dispersion of nanoparticles in the phase (CLC). The emission data presented in Figure 5 clearly demonstrate the results. QD films placed outside the cavity produced minimal detected emission in leg 1 as only one handedness of light was able to pass through the cavity. A comparable concentration of QDs dispersed in the cavity produced emission and clear fringes characteristic of cavity coupling.

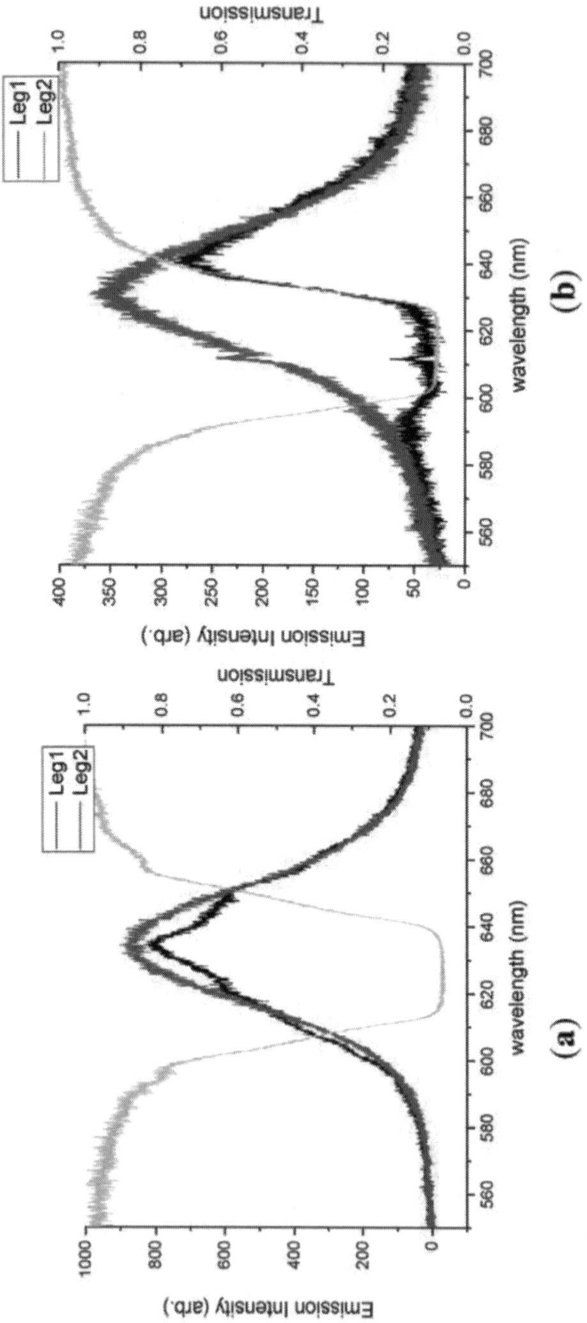

Figure 5. Photoluminescence spectra for 0.02 wt% ODA-QDs. This data compares (**a**) QD dispersion in the cholesteric phase with (**b**) QDs located beneath the cholesteric film.

4. CONCLUSIONS

In this paper, we investigated quantum dots dispersed in a cholesteric photonic cavity for potential liquid crystal laser applications. We primarily compared the effects of changing the QD ligand to a mesogenic molecule recently studied by our group. In addition, we also looked at locating the QDs outside of the cavity compared with the composite materials.

We found that QDs used in the study produced cavity coupling signatures, but, at fairly low loading concentrations, they also suffered from significant aggregation. The LC–QDs demonstrated reduced aggregation, although we should note that it is still not clear what the role of this aggregation is on device performance. It may be that as long as the bulk cholesteric texture is not too disrupted full dispersion is not necessary for a useful device. Interestingly, we noted that strong coupling effects, as measured with a 600 nm spot size, where spatially dependent indicating that hitting the material close to an aggregate produced stronger effects due to a larger localized emitter concentration.

No significant effects on the emission spectrum were observed between the two particle types, despite the fact that QDs in the ODA–QD clusters are closer together (~7.6 nm [13]) compared with LC–QDs (~11 nm [18]). Localizing the QD film outside the cavity produced detrimental effects on performance, as the QD emission detected in leg 1 was reflected and did not travel through the CLC.

From these results, we speculate that very well dispersed QDs are not necessary to produce strong cavity coupling in a cholesteric liquid crystal device for lasing applications and that aggregation may be advantageous for increasing signal strength. Such results support recent observations of lasing in a similar device with very high QD content (up to 8 wt%) [21] in which aggregation and non-uniform QD distribution is likely. We propose that controllably assembled QD aggregates such as the QD shells [14] and membranes shown in Figure 3 may find application in photonic devices, provided the host phase alignment is not significantly disrupted by the embedded emitting structures.

ACKNOWLEDGMENTS

This work was supported in part by a grant from the nanoBIO node of the National Science Foundation (ECC-1227034) and the Materials Research Laboratory (University of Illinois at Urbana-Champaign). The authors would also like to acknowledge generous funding from National Science

Foundation grants DMR 0852791, DMR 1056860, and from the University of California, Merced Graduate and Research Council.

AUTHOR CONTRIBUTIONS

Linda S. Hirst, Sayantani Ghosh and Jason E. Hein conceived and designed the experiments; Andrea L. Rodarte and Fredy Cisneros performed the experiments; Andrea L. Rodarte analyzed the data; Linda S. Hirst, Sayantani Ghosh and Jason E. Hein contributed reagents/materials/analysis tools; Linda S. Hirst wrote the paper.

CONFLICT OF INTEREST

The authors declare no conflict of interest.

REFERENCES

1 Overgaag, K.; Evers, W.; De Nijs, B.; Koole, R.; Meeld-ijk, J.; Vanmaekelbergh, D. Binary Superlattices of PbSe and CdSe Nanocrystals. *J. Am. Chem. Soc.* **2008**, *130*, 7833–7835

2 Tao, A.; Sinsermsuksakul, P.; Yang, P. Tunable plasmonic lattices of silver nanocrystals. *Nat. Nanotechnol.* **2007**, *2*, 435–440.

3 Shevchenko, E.V.; Talapin, D.V.; Kotov, N.A.; O'Brien, S.; Murray, C.B. Structural diversity in binary nanoparticle superlattices. *Nature* **2006**, *439*, 55–59.

4 Romo-Herrera, J.M.; Alvarez-Puebla, R.A.; Liz-Marzan, L.M. Controlled assembly of plasmonic colloidal nanoparticle clusters. *Nanoscale* **2011**, *3*, 1304–1315.

5 Ofir, Y.; Samanta, B.; Rotello, V.M. Polymer and biopolymer mediated self-assembly of gold nanoparticles. *Chem. Soc. Rev.* **2008**, *37*, 1814–1825.

6 Cheng, W.; Campolongo, M.J.; Cha, J.J.; Tan, S.J.; Umbach, C.C.; Muller, D.A.; Luo, D. Free-standing nanoparticle superlattice sheets controlled by DNA. *Nat. Mater.* **2009**, *8*, 519–525.

7 Srivastava, S.; Frankamp, B.L.; Rotello, V.M. Modulation of the Interparticle Spacing and Optical Behavior of Nanoparticle

Ensembles using a Single Protein Spacer. *Chem. Mater.* **2005**, *17*, 487–490.

8 Shenton, W.; Davis, S.A.; Mann, S. Directed Self-Assembly of Nanoparticles into Macroscopic Materials Using Antibody–Antigen Recognition. *Adv. Mater.* **1999**, *11*, 449–452.

9 Bockstaller, M.R.; Lapetnikov, Y.; Margel, S.; Thomas, E.L. Size-Selective Organization of Enthalpic Compatibilized Nanocrystals in Ternary Block Copolymer/Particle Mixtures. *J. Am. Chem. Soc.* **2003**, *125*, 5276–5277.

10 Cui, H.; Chen, Z.; Zhong, S.; Wooley, K.L.; Pochan, D.J. Block Copolymer Assembly via Kinetic Control. *Science* **2007**,*317*, 647–650.

11 He, J.; Liu, Y.; Babu, T.; Wei, Z.; Nie, Z. Self-Assembly of Inorganic Nanoparticle Vesicles and Tubules Driven by Tethered Linear Block Copolymers. *J. Am. Chem. Soc.* **2012**, *134*, 11342–11345.

12 Prodanov, M.F.; Pogorelova, N.V.; Kryshtal, A.P.; Klymchenko, A.S.; Mely, Y.; Semynozhenko, V.P.; Krivoshey, A.I.; Reznikov, Y.A.; Yarmolenko, S.N.; Goodby, J.W.; *et al.* Thermodynamically Stable Dispersions of Quantum Dots in a Nematic Liquid Crystal. *Langmuir* **2013**, *29*, 9301–9309.

13 Rodarte, A.L.; Pandolfi, R.J.; Ghosh, S.; Hirst, L.S. Quantum dot/liquid crystal composite materials: Self-assembly driven by liquid crystal phase transition templating. *J. Mater. Chem. C* **2013**, *1*, 5527–5532.

14 Rodarte, A.L.; Cao, B.H.; Panesar, H.; Pandolfi, R.J.; Quint, M.; Edwards, L.; Ghosh, S.; Hein, J.E.; Hirst, L.S. Self-assembled nanoparticle micro-shells templated by liquid crystal sorting. *Soft Matter* **2015**, *11*, 1701–1707.

15 Coles, H.; Morris, S. Liquid Crystal Lasers. *Nat. Photonics* **2010**, *4*, 676–685.

16 Lukishova, S.G.; Bissell, L.J.; Winkler, J.; Stroud, C.R. Resonance in quantum dot fluorescence in a photonic-bandgap liquid crystal host. *Opt. Lett.* **2012**, *37*, 1259–1261.

17 Lukishova, S.G. Liquid crystals under two extremes: (1) High-power laser irradiation, and (2) Single-photon level.*Mol. Cryst. Liq. Cryst.* **2012**, *559*, 127–157.

18 Rodarte, A.L.; Nuno, Z.S.; Cao, B.H.; Pandolfi, R.J.; Quint, M.; Ghosh, S.; Hein, J.E.; Hirst, L.S. Tuning quantum dot organization in liquid crystal for robust photonics applications. *Chem. Phys. Chem.* **2014**, *15*, 1413–1421.

19 Rodarte, A.L.; Grey, C.; Hirst, L.S.; Ghosh, S. Spectral and polarization modulation of quantum dot emission in a one-dimensional liquid crystal photonic cavity. *Phys. Rev. B* **2012**, *85*.

20 Rodarte, A.L.; Shcherbatyuk, G.; Shcherbatyuk, L.; Hirst, L.S.; Ghosh, S. Dynamics of spontaneous emission of quantum dots in a one-dimensional cholesteric liquid crystal photonic cavity. *RSC Adv.* **2012**, *2*, 12759–12763.

21 Chen, L.J.; Lin, J.D.; Lee, C.R. An optically stable and tunable quantum dot nanocrystal-embedded cholesteric liquid crystal composite laser. *J. Mater. Chem. C* **2014**, *2*, 4388–4394.

CHAPTER 7

Low-Temperature Bonding for Silicon-Based Micro-Optical Systems

Yiheng Qin, Matiar M.R. Howlader * and M. Jamal Deen

Department of Electrical and Computer Engineering,
McMaster University, Hamilton, ON, L8S 4K1, Canada

ABSTRACT

Silicon-based integrated systems are actively pursued for sensing and imaging applications. A major challenge to realize highly sensitive systems is the integration of electronic, optical, mechanical and fluidic, all on a common platform. Further, the interface quality between the tiny optoelectronic structures and the substrate for alignment and coupling of the signals significantly impacts the system's performance. These systems also have to be low-cost, densely integrated and compatible with current and future mainstream technologies for electronic-photonic integration. To address these issues, proper selection of the fabrication, integration and assembly technologies is needed. In this paper, wafer level bonding with advanced features such as surface activation and passive alignment for vertical electrical interconnections are identified as candidate technologies to integrate different electronics, optical and photonic components. Surface activated bonding, superior to other assembly methods, enables low-temperature nanoscaled component integration with high alignment accuracy, low electrical loss and high transparency of the interface. These features are preferred for the hybrid integration of silicon-based micro-opto-electronic systems. In future, new materials and assembly technologies may emerge to enhance the performance of these micro systems and reduce their cost. The article is a detailed review of bonding techniques for electronic, optical and photonic components in silicon-based systems.

KEYWORDS

hybrid integration; low-temperature bonding; silicon photonics; optical detection systems; surface activated bonding

1. INTRODUCTION

Silicon (Si)-based micro-optical systems are of great importance for environmental and healthcare applications. Such systems, which can be highly integrated, enable fast and efficient signal acquisition and data processing. In addition, the fabrication of Si-based devices is compatible with parallel processing which can potentially lower the cost. One practical example of such system is the early diagnosis and monitoring of human diseases using autofluoresnce or fluorescence lifetime imaging [1]. Another example is a bioanalytical sensing system that allows detection and monitoring of Escherichia coli cell in food and water [2]. One of the challenges for such systems is to integrate optical, electrical, and fluidic components on a single platform. In addition, these systems should be portable, easy-to-use, power efficient, and low-cost. Therefore, fabrication, integration and assembly of components or modules in a small form factor are required, but this is not easy to realize in practice.

One of the challenges to realize the highly-integrated systems is the electrical interfacing among the components and modules. It is required for the transmission and distribution of electrical power and data signal. The data rate for the integrated optical systems is expected to be 80 Tb/s to 780 Tb/s by 2022 [3]. In such high-speed systems including electronic components, the geometries and materials for electrical interconnects must be carefully designed to optimize their high frequency performance [4]. In general, interconnects with smaller dimensions can help to reduce electrical resistance, capacitance and/or inductance to improve the high-frequency performance.

Another challenge for systems integration is the optical interfacing among optical components. It is required for transmitting optical signals. The quality of optical interfaces is determined by the extent of matching between the electromagnetic fields of the optical sources, detectors and waveguides. A high coupling efficiency is required for transferring optical energy efficiently across the interface between the coupled components. For example, since the relative position offsets between the optical source and detector control coupling loss [5], their alignment accuracy during integration is critical. Briefly, passive alignment using groove structures is more practical than active alignment for commercial micro-optical systems.

Currently, wafer bonding technologies are used to attach two or more substrates with different functional components, and are being used for the electrical interconnection and optical coupling [6]. These technologies provide bonding of devices with small dimensions and short interconnects throughout the wafers. Also, these technologies offer high throughput and low cost, because all components on the same wafer are processed in parallel [7]. Therefore, the total cost of the parallel-processed product is lower than that of the products assembled discretely.

Bonding of Si wafers with direct band gap materials (III-V materials) offers the possibility to fabricate Si-based photonic devices such as laser diodes [8]. In this case, Si, as a substrate, guides the optical signals generated from the III-V materials, which have high optical gains. Also, it is important to integrate the optical devices with complementary metal-oxide-semiconductor (CMOS) circuits to read, process, store, and transmit obtained optical signals.

For the integrated optoelectronic systems, the development of suitable bonding technologies for Si wafers is one of the critical steps. Unfortunately, current bonding technologies, such as fusion bonding [9] or thermo-compression bonding [10], do not provide the following properties at the same time: (1) high bonding strength, (2) high electrical conductivity, (3) hermetic sealing, (4) high transparency of the bonded interface, and (5) low thermomechanical stress in the bonded components. Also, since many optoelectronic systems operate at a temperature lower than the temperatures used for fusion and thermo-compression bonding, the mismatch of coefficient of thermal expansion (CTE) of different materials would introduce alignment error during bonding. The misaligned components would result in degraded system performance. Moreover, the CTE mismatch between metal coating and Si-based micromirrors causes unwanted mirror wrapping and bowing [11,12]. Therefore, low-temperature bonding for the integration and assembly of optical detection systems is required. Indirect bonding uses intermediate layers such as solders, metals, and adhesives. In contrast, direct bonding uses atomic forces between the surfaces of mating pairs. Surface activated bonding (SAB) is an example of the direct wafer bonding [13]. The SAB technologies do not require high temperature, high pressure, and extra additives to provide bonding interface with high bond strength, high electrical conductivity and low optical loss [13,14,15,16,17,18,19,20,21,22,23,24,25,26,27,28].

There are optically sensitive materials such as carbon nanotubes (CNTs) [29], graphene [30], molybdenum disulfide (MoS_2) [31], tungsten disulfide (WS_2) [32], and organic polymers [33] to be integrated with Si-based optical systems. This integration offers highly-sensitive and power efficient systems because of the unique properties of the optical materials such as short response time and high photoresponsivity. Thus, advanced assembly methods for new materials are required to be developed. Monolithic integration and heterogeneous integration contribute to the advanced assembly concepts to further enhance the performance of devices and systems. Note that the targeted assembly cost for Si-based optical detection systems is majority of the overall cost [34]. This cost may be increased when new materials are to be integrated with Si.

In this article, we review in detail the fabrication, integration, and assembly of Si-based micro-optical systems using wafer bonding at low temperatures (below 450 °C). We focus on the electrical and optical interfaces in such systems. In Section 2, we present basic concepts and technologies for electrical and optical interfaces in micro-optical systems. In Section 3, we describe low-temperature wafer bonding technologies in detail, with emphasize on the SAB technologies. Different aspects of direct bonding without intermediate layers and indirect bonding with intermediate layers are explained. In Section 4, we identify different aspects, such as new materials, technologies, and cost, for future Si-based micro-optical systems.

2. ELECTRICAL AND OPTICAL INTERFACES IN MICRO-OPTICAL SYSTEMS

2.1. Electrical and Optical Interfaces

A typical microsystem requires electrical interconnects and optical coupling for the communication among neighboring components, as shown in Figure 1. This system consists of electronic, optical and photonic devices with packages. The electrical and optical interfaces help to attain system functions, performance, stability, and reliability. For example, the intensity of electrical signals in the system is affected by resistance of interconnects. Thus, the location, dimensions, and materials of electrical interconnects have to be well designed. In addition, the intensity of optical signals received by the detectors impacts the sensitivity and detection limit of the system. Therefore, precise alignment among the optical components is indispensable to ensure efficient signal transmission. For this purpose, the electrical and optical connection in a Si-based microsystem will be discussed in detail in this subsection.

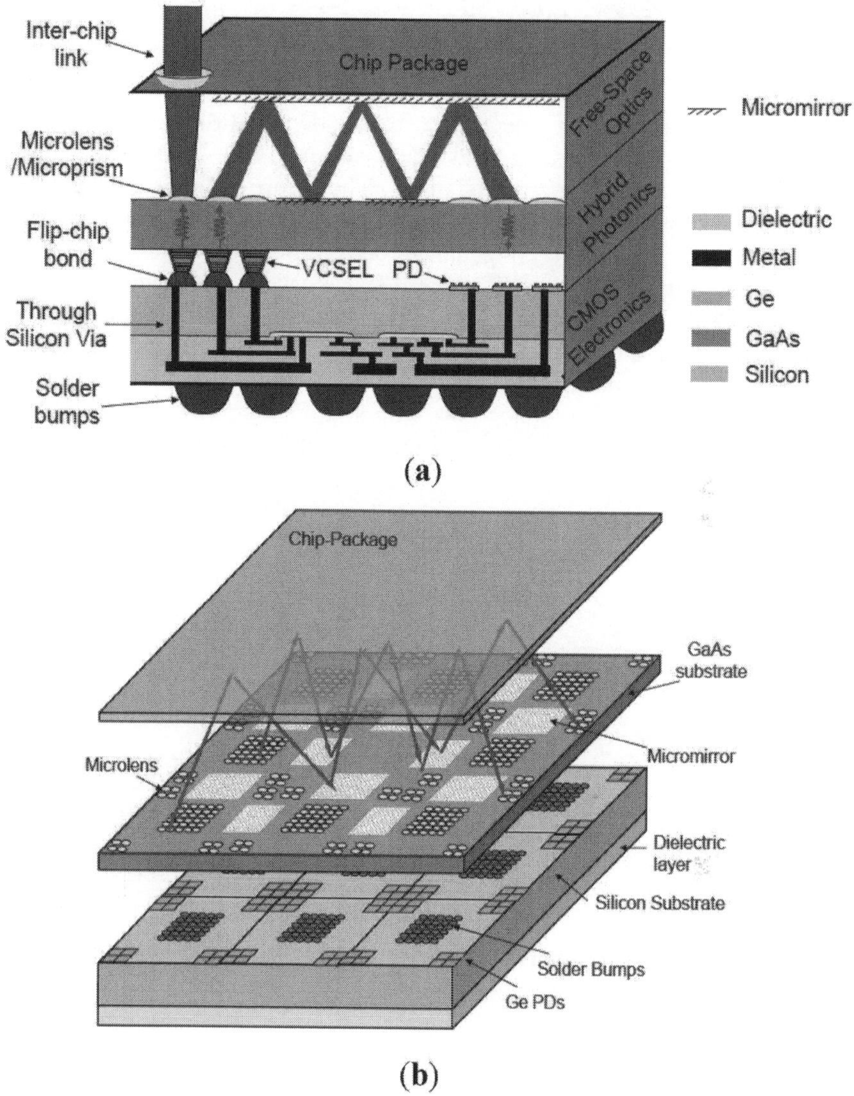

Figure 1. (**a**) Cross-sectional and (**b**) three-dimensional view of a typical microsystem that consists of electronic (*i.e.*, complementary metal-oxide-semiconductor (CMOS) transceivers and amplifiers), optic (*i.e.*, lenses and mirrors) and photonic (lasers and photodetectors) components with packages [35] (Reproduced with permission from Ciftcioglu *et al.*, *Opt. Express*; published by OSA, 2012).

2.1.1. Electrical Interconnects

Electrical interconnects are necessary for transmitting power and signals in many systems with electronic components. Propagation, delay, attenuation, noise, crosstalk and the rise and fall times have been observed in high frequency signals [36]. The electrical interconnects in a system with optical components should possess small time constant with low resistance. Also, the impedance between the signal sources and interconnects requires being matched [37]. Vertical interconnections, as one type of electrical interconnects, offer reduced resistance and parasitic parameters, and the device can also be densely integrated due to the reduced in-plane space [38]. Micro bumps and through-hole vias are needed for vertical interconnections with small foot-print. The schematic of vertical interconnections using bumps and through-hole vias is shown in Figure 2.

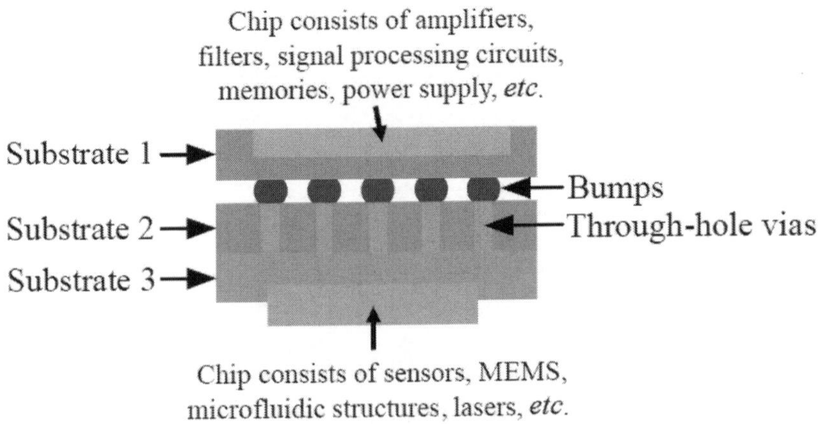

Figure 2. Schematic of vertical interconnections between bumps and through-hole vias for the components shown in Figure 1.

Bumps are formed in the last few steps of fabricating a component before it (substrate 1) is assembled onto target substrates (substrates 2 and 3 in Figure 2). Conducting adhesives and metals/alloys are generally being used to make bumps. Bumps made of conducting adhesives are not discussed here due to their low conductivity, although, in some studies, nanoparticles are added in polymer matrixes to enhance the conductivity, the conductivity is still lower than that of metal [39]. Bumps made of metals and eutectic alloys are mostly used because of their good electrical and thermal conductivity that reduces electrical power consumption, and increases heat dissipation. By controlling the size of the metal/alloy bumps, the capacitance and inductance of the interconnections could be minimized [40].

The diversity of the bump materials and substrates causes compatibility issues during heating steps in assembly. To address this issue, soft alloys, for example tin-lead (Sn-Pb), have been applied to reduce stress in the assembled devices and systems [41]. However, the European union regulations on the restriction of hazardous substances in electrical and electronic equipment stimulated the development of lead-free soldering and bumping technologies [42]. Lead-free alloys for bump materials include Sn-silver (Ag), Sn-Ag-copper (Cu), Cu-Sn, gold (Au)-Sn and Au-indium (In). Metallic bumps such as In and Au have also been used. Indium bumps are required for optical detection systems because of its low stress [43]. The low stress is attributed to its high ductility and low Young's modulus. This property is important for structures having mismatches in coefficient of thermal expansion (CTE). Besides, Gold micro stud bumps have been used to connect Si substrates with photonics materials such as $LiNbO_3$ [26].

Vacuum deposition and electroplating techniques are used to fabricate bumps, but the electroplating approach is now preferred because of its low cost. Electroplating has been demonstrated for making bumps with pitches of 40 μm [43,44], or even smaller [45]. Furthermore, special techniques in electroplating such as pulse plating or pulse reverse plating may improve the uniformity and consistency of the bump height throughout the wafer.

In addition to bumps, through-hole vias are required in vertical interconnections. Large-scale optical detector arrays were assembled in the form of sensing surface facing up, and the electrical signals were extracted from the back side of the chip using through-hole vias [46]. This approach does not sacrifice the sensing surface area of the front side and maximizes the fill-factor of the sensing component. The vias can be vertically fabricated by etching, filling, and grinding through the substrates. The lengths of the interconnections are identical to the substrate thicknesses, thus this interconnection method can achieve high conductivity. Also, low parasitic capacitance, high breakdown voltage, and small leakage current can be realized when proper passivation techniques are used [46].

System integration using through-hole vias could be divided into three categories [13,47]: via-first, via-middle and via-last. In via-first processes, the components and filled vias are fabricated on different substrates before wafer bonding. During wafer bonding, structures to be aligned are close to bottom-level components, which are densely distributed and small in dimensions. Therefore, precise alignment is required but challenging. Also, high precision alignment is time-consuming. It reduces yield, increases processing complexity and cost. Similarly, via-middle process is carried out after the device fabrication but before metallization steps. The small and dense structures make the alignment challenging. On the other hand, the via-last processes are done at the last step in the assembly. Therefore, via-last approaches do not require precise wafer-to-wafer alignment during wafer bonding, thus reducing the processing complexity and cost.

2.1.2. Optical Coupling

A microsystem with optical components uses optical signals from optical sources, waveguides for signal transmission, and often detectors for signal detection (Figure 1). These components may be discrete or integrated. In the case of discrete components, optical coupling is realized between them. For practical applications, for example for detecting weak signals such as biological fluorescence, high coupling efficiency is one of the most important parameters of the interface. The coupling efficiency is affected by the position misalignment between optical sources, waveguides, and detectors, the numerical aperture mismatch of waveguides, and the refractive index mismatch between coupled components. Among them, misalignment between optical sources and detectors has a significant impact on coupling efficiency [5]. Thus, the optical source and detector have to be precisely aligned. Although the alignment tolerances of large diameter fibers may be relaxed for short distance optical interconnect systems [48], they are very tight for both lateral coupling (± 0.3 μm for 1 dB decrease in coupling efficiency) and vertical coupling (± 2 μm for 1 dB decrease in coupling efficiency) of small-size integrated systems. There are mainly two schemes to achieve optical alignment [49]: active alignment and passive alignment.

Active alignment (Figure 3a) has been demonstrated in an indium phosphide (InP) photonic device for an integrated wavelength division multiplexing transceiver [50]. In this approach, a lensed fiber was precisely aligned to the optical output using closed-loop optimization. It is a time consuming and expensive process with low throughput, especially when multiple fibers are to be connected.

Passive alignment (Figure 3b) is a promising strategy because of its short alignment time, high throughput, low cost, compatibility with micro-structures, and multichannel compatibility [26]. It may also eliminate the need for equalization circuits and transmission lines used in active alignment. In this approach, optically connected components can be densely assembled on one substrate using this approach [51,52]. However, the performance of systems assembled by passive alignment varies from each other. This variation is due to the reduced alignment accuracy resulting from the accumulated mechanical tolerance of the alignment machine [38].

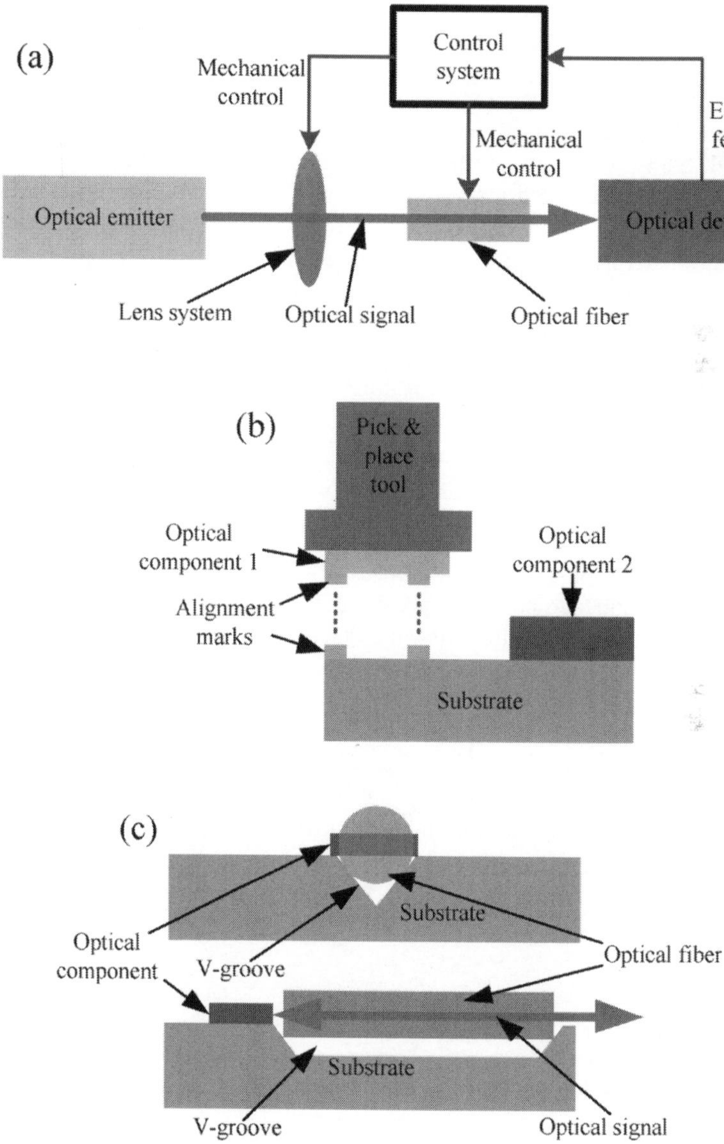

Figure 3. Schematic of optical alignment methods. (**a**) Close-loop active alignment; (**b**) Passive alignment using alignment marks; (**c**) Self-alignment using V-groove.

Several methods of low process complicity were developed to improve passive alignment accuracy. In the first method [53], the challenge in the planar alignment due to height differences of the optical components has been addressed. Firstly, the optical source and detector are mounted on a planar substrate with a mirror parallel to the substrate. When the optical signal is generated from the source, it is reflected by the mirror and finally recorded by the detector. By this process, coupling of optical signals in the vertical orientation through the edge of devices can be eliminated. In another study, the high temperature issue in alignment was addressed using low-temperature bonding to avoid re-melting of the bumps in highly integrated micro-optical systems. This method gets rid of the surface tensions of melted bumps (liquid state) that may displace connected structures [54,55]. Low-temperature integration technology based on passive optical alignment using SAB has been demonstrated to minimize alignment offset induced by mismatch of materials' CTE [56]. Germanium (Ge) [56] and $LiNbO_3$ [8] were bonded to Si by this method for the fabrication of micro-laser sources and micro-optical sensors. Finally, precise alignment through glass was demonstrated for optoelectronic modules. The benefits of this method were to utilize the dimensional stability under thermal load, transparency, and CTE matching to Si [57].

Besides active and passive alignment, a self-alignment concept for high-quality optical interface has been presented (Figure 3c) [38,58]. Firstly, a groove structure is fabricated on the substrate to hold the waveguide according to the position of the optical source/detector. The dimensions of the groove are designed for fitting the waveguide (optical fiber in this example) with a certain diameter. Then, the optical fiber can be placed and fixed inside the groove. Finally, the alignment between the waveguide and the source/detector is automatically achieved by the specified position of the waveguide in the groove. The grooves could be a V-shape trench prepared by Si wet etching or a rectangular trench realized by Si deep reactive ion etching (RIE) [59]. In another study, low-cost devices like datacom modules have been fabricated using similar MEMS technologies [60].

2.2. Wafer Bonding for Electrical and Optical Interfaces

A schematic representation of a Si-based optical sensing microsystem is shown in Figure 4. It consists of an optical source, waveguides, gratings, detectors, electronics, a power supply and a wireless unit. All these components should be integrated on a single common platform for miniaturization. The electrical interfaces should possess interconnects with criteria of low loss, high conductivity, and high signal transmission rate. The optical components should be precisely aligned to obtain maximum coupling efficiency across the interfaces. Conventional approaches, such as wire bonding and lead frame assembly, require large footprint of device because of the long wires and housing lids, which introduce extra cost. Wafer bonding is an improved technology that meets requirements of downscaling, increasing integration density and lowering fabrication cost [61].

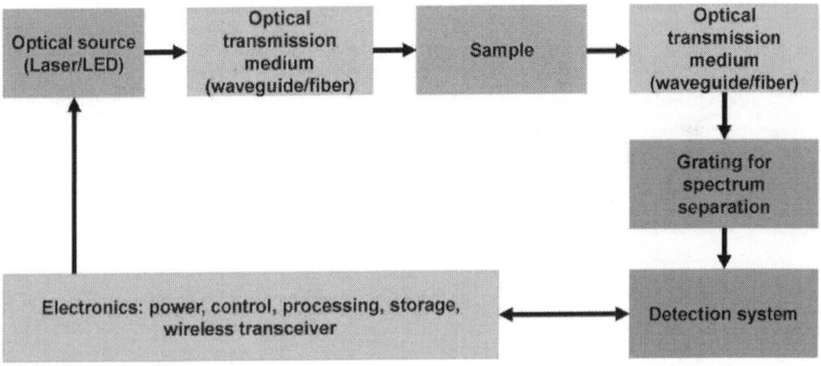

Figure 4. Schematic of an optical sensing microsystem using a Si-based platform. The sample could be a fluidic system with biological species, a chamber with gas, or a solution with target molecules.

Creating electrical interfaces by wafer bonding usually incorporates the mechanical interconnection of conductive materials (bumps) on two wafers. The dimensions of such interconnections are minimized to reduce the capacitance and leakage current at the interface. Specifically, stud bump bonding is a high-speed, modified wire bonding process that is compatible with wafer-level processing and can be used for electrical interconnection. Different stud bumping processes are available. Gold stud bumping has been reported with a bump pitch less than 50 μm when wires with 15–33 μm diameters are used [37]. An example of Au microstud bumps have been used to connect Si substrates with photonics materials such as $LiNbO_3$ is shown in Figure 5 [26]. One example of Cu TSV bonded with Au stud bump is also shown in Figure 5 [62]. Also, Cu stud bumping has been demonstrated on Al metallization in CMOS technologies [63]. Cu is preferred over Au stud bumping because of the slower growth of Cu-Al intermetallic phases and the lower cost of the material [64]. However, when implementing Cu as the bumping material, the oxidation of Cu surfaces and the requirement for large bonding forces should be taken into consideration.

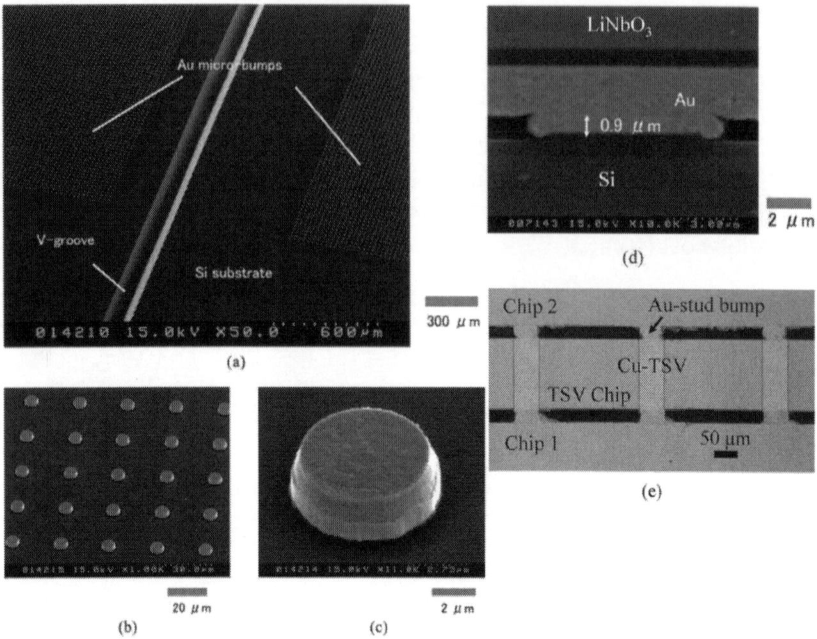

Figure 5. (**a**) SEM image of a Si substrate with a V-groove and Au micro-bumps; (**b**) SEM image of Au micro-bumps; (**c**) Magnification of (b); (**d**) Cross-sectional SEM image of an Au micro-bump on a Si substrate after bonding with a $LiNbO_3$ substrate [26] (Reproduced with permission from Takigawa *et al.*, *IEEE J. Sel. Top. Quantum Electron.*; published by IEEE, 2011); (**e**) Optical image of cross-sectional view of vertically integrated system between Au-stud bumps and Cu-TSVs [62] (Reproduced with permission from Howlader *et al.*, *J. Vac. Sci. Technol. A*; published by AIP, 2011).

In terms of optical interfaces, one study has shown that hermetic sealing is necessary for high performance optical detection systems [65]. For example, infrared detectors such as microbolometers must be packaged inside vacuum cavities (normally with a getter inside the cavity for the level of low pressure needed for microbolometers). The minimization of thermal conduction through controlling residual gases in the well-controlled package environment reduces the loss of thermoelectricity of the detectors. Here, this wafer level vacuum assembly is carried out by bonding of wafers with cavities to house the detectors in the hermetically sealed vacuum [66]. Thus, wafer bonding provides the system with a low-loss optical interface between the system and external environment.

In addition to the housing technique, wafer bonding can also be used for the fabrication of optical devices with high-quality interfaces between optical sources/detectors and waveguides. Current Si photonic technologies combine the efficient light emission benefit of III-V materials (such as InP) with the low losses and high index contrast waveguiding benefit of Si [67,68]. An approach has been developed to combine these advantages. In one attempt, III-V materials are directly deposited on Si substrate. However, this approach failed due to the lattice mismatch between the deposited materials and the underlying Si substrate. This mismatch was because of the large differences between the deposition temperature and the temperature for using the devices [69,70]. An alternative approach using wafer bonding was developed to address this integration problem by directly bond two wafers at low temperatures [63,71]. The low-temperature direct wafer bonding technologies allow for reducing the lattice mismatch and results in high-quality integration with void-free bonded interfaces. For example, a III-V wafer is bonded to a Si-on-insulator (SOI) wafer with optical waveguides (Figure 6). The III-V quantum wells in the bonded wafer provide optical gain by electrical injection of electrons and holes across a pn junction. The small index difference between Si and III-V materials provides a degree of freedom in manipulating modal confinement in Si and III-V, which is realized by adjusting the III-V layer thickness and SOI waveguide dimension. To decrease the cost in this process, a die-to-wafer bonding process was employed [64]. Also, an InP microdisk laser was bonded to a Si wafer with waveguide by hydrophilic wafer bonding at room temperature [40]. By post-annealing at 200 °C, a good optical interface between the InP laser and Si waveguide was realized.

In addition to III-V compounds, single crystalline Ge is also an important material in high sensitivity and high-speed optical detectors at wavelengths up to 1.6 μm. Low-temperature wafer bonding has been observed to create junctions between Ge and Si for the detectors [73]. This method provides less dislocation defects and better crystal quality than in the epitaxially grown Ge which has a lattice mismatch as large as 4.2%. The junction resistance of the two bonded Ge wafers was similar to their bulk resistance, showing a high-performance bonding interface in terms of electrical conduction [56]. Moreover, Ge optical detectors with Si waveguides have been fabricated using Si-Ge wafer bonding. The resulting dark current density was 4 nA/μm^2 and the quantum efficiency was over 90% [73].

Lastly, another advantage of wafer bonding is that different types of components (electronics, optical units, packages) can be developed, processed, and optimized separately before bonding [44]. Many bonding technologies used in the fabrication and assembly of Si-based micro-optical systems use high temperature, high pressure, vacuum bonding chamber, and thick adhesives [74]. These technologies may degrade the materials on the substrates (e.g., polymer), increase process and assembly costs, require complicated equipment (e.g., ultra-high vacuum bonder for vacuum seal), and sometimes be incompatible with biological specimens.

Thus, high-performance, reliable, CMOS compatible and cost-effective wafer bonding technologies using reduced external agents (such as heat, force, and adhesive) have to be developed for emerging sensing applications in health and environmental monitoring. A detailed review regarding these approaches is given in the next section.

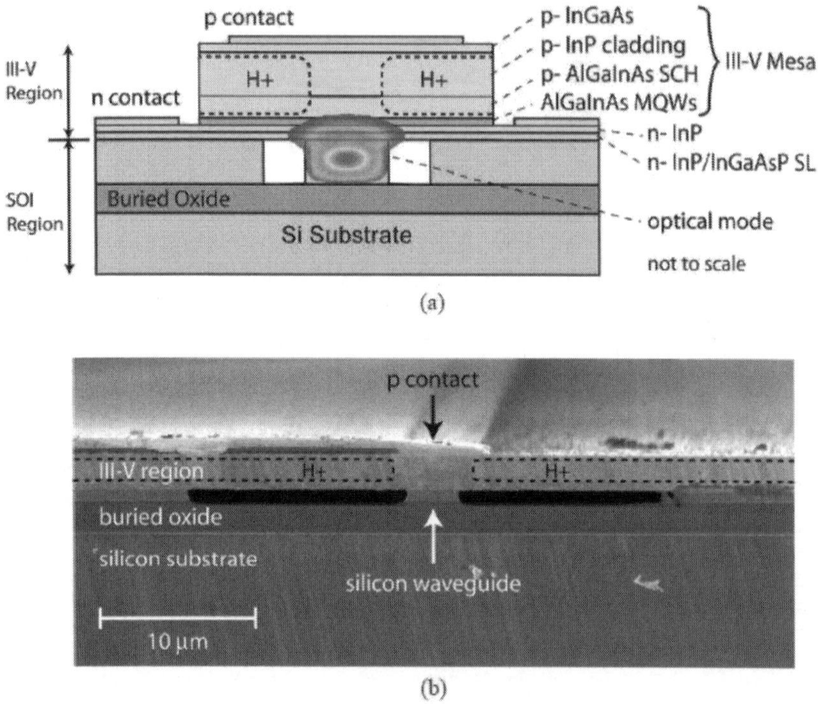

Figure 6. (**a**) Cross-sectional schematic of hybrid Si evanescent platform; (**b**) SEM image of hybrid Si evanescent platform [72] (Reproduced with permission from Fang *et al.*, *Opt. Express*; published by OSA, 2006).

3. LOW-TEMPERATURE WAFER BONDING TECHNOLOGIES

In the previous section, wafer bonding was described as an important technology to fabricate and assemble Si-based micro-optical systems. In order to obtain high performance electrical and optical interfaces, low-temperature wafer bonding is recommended. Thermal treatment at low temperatures minimizes the CTE difference induced alignment offset and stress between materials, thus increasing the optical coupling efficiency and the reliability of the interconnections [61].

The fabrication of Si-based micro-optical systems requires high-quality bonded interfaces. The criteria include but are not limited to [16]:

(1) High bond strength, which ensures the mechanical stability and reliability of the assembled optoelectronic components, and offers stability in the optical alignment.

(2) High electrical and thermal conductivity to reduce the energy loss in the electronic components and the high thermal conductivity enables efficient thermal management of the working device.

(3) High hermeticity of the seal to provide the desired working environment for certain devices (such as infrared detectors), thus improving the performance of the system.

(4) High alignment accuracy, which is essential in optical devices for a high optical coupling efficiency.

(5) Excellent optical transparency to reduce the reflection of incident light.

To achieve the aforementioned requirements of the bonded interfaces for the optical detection systems, low-temperature (below 450 °C, which is CMOS compatible [75]) wafer bonding processes are needed. They are divided into two categories:

(1) Direct bonding, for example anodic bonding, SAB, hydrophilic bonding, and plasma assisted wafer bonding; and

(2) Indirect bonding with intermediate layers, such as adhesive bonding, solder bonding, and SAB with nanoadhesion layers.

In this section, different types of low-temperature wafer bonding are discussed, with emphasizing on the SAB technologies. Their advantages will be highlighted and their disadvantages will be described.

3.1. Direct Bonding without Intermediate Layers

In optical systems, high transparency of optical interfaces is important to maximize optical coupling efficiency. Therefore, intermediate materials should be minimized at the interface of the coupled structures to avoid attenuation of optical signals and variation of refractive indexes. To address this issue, low-temperature direct wafer bonding technologies were developed.

3.1.1. Anodic Bonding

In anodic bonding method, a silicon wafer is bonded with an alkaline (such as sodium) rich glass substrate under the application of 200 V to ~1000 V in the typical temperature ranges of 200 °C to 400 °C [76,77], as shown in Figure 7a. However, the strict requirements may change considerably the volume of glass, causing bowing or warping of the bonded wafer pair. Such mechanical deformation is problematic for optical coupling [78]. Moreover, oxygen anions are generated during anodic bonding at the Si/glass interface. They may oxidize the electrical interconnect metals, resulting in high resistance.

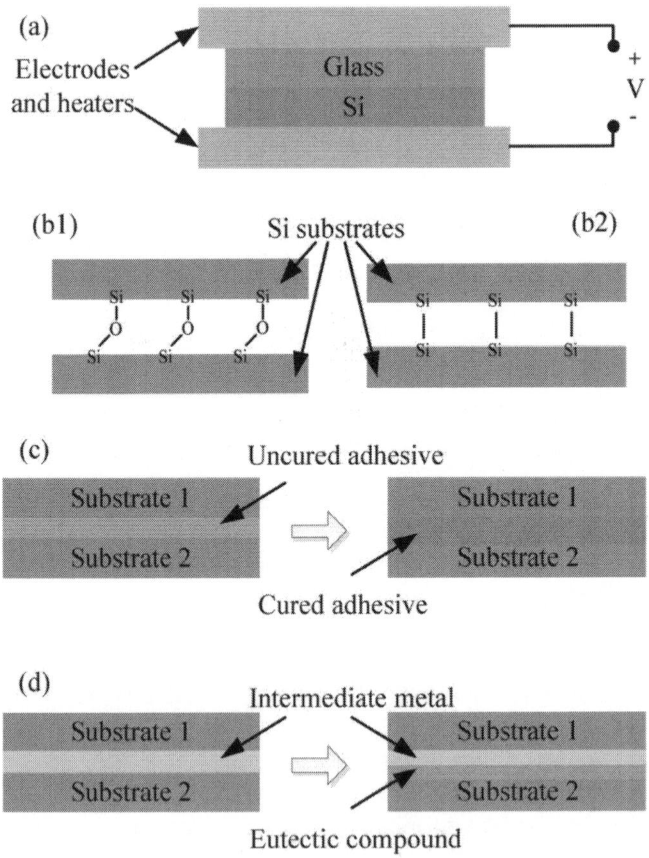

Figure 7. Schematic of: (**a**) Anodic bonding technique to bond glass and Si substrates; (**b**) Hydrophilic and hydrophobic bonding; (**c**) Adhesive bonding; (**d**) Eutectic bonding. Components being used in the substrates for practical applications are not shown in the figure.

In the integration of micro light-emitting diodes (LEDs) or optical detectors, anodic bonding was used to form hermetic optical caps [79] (Figure 8). First, cavities were etched in a Si substrate, followed by mounting optical components such as detectors or LEDs inside the cavities. Then, anodic bonding between the Si substrate and a borosilicate glass substrate was carried out at a maximum electrical field of 20 V/μm and a temperature of 390 °C for 10 min. The fabricated optical windows were smaller than 0.75 mm × 0.75 mm with a yield higher than 95%. Except for the wide applications in fabricating optical caps [80], anodic bonding was also used for building a vacuum tight vapor cell for optogalvanic spectroscopies [81], microlens arrays [82], 3D microlens scanners [83,84], microfluidic-based optical detection systems [85], micro-optical magnetometers [86], and micro-optical choppers for Raman spectroscopies [87].

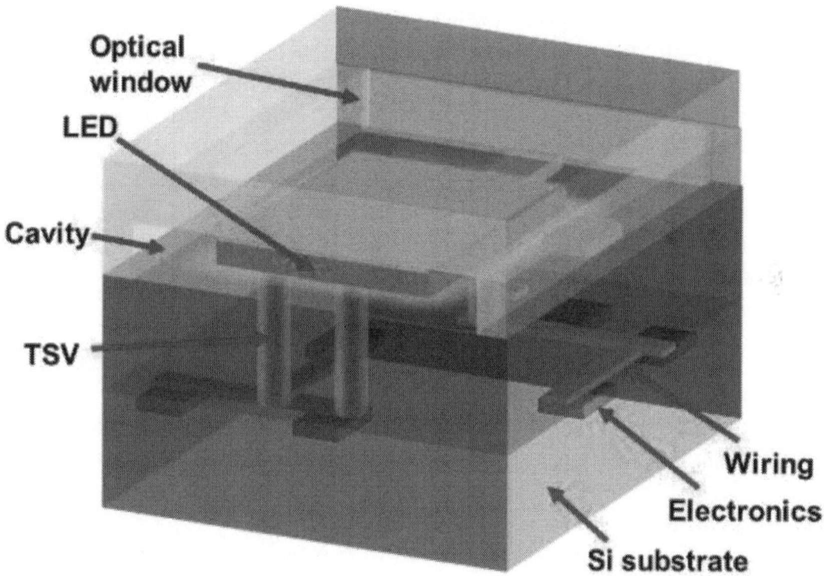

Figure 8. Schematic of a 3D-integrated light-emitting diode (LED) with anodically bonded optical window.

3.1.2. Hydrophilic and Hydrophobic Bonding

Anodic bonding requires an alkaline rich glass wafer in the bonding pair. Also, it needs the application of high voltage during bonding. Therefore, hydrophilic and hydrophobic bonding technologies were introduced to bond two pieces of Si wafers. The hydrophobic bonding is resulted from the reactions between chemical terminations on the substrates. For Si substrates, terminations of Si-F and Si-H are formed during surface treatment such as dipping in hydrofluoric acid [88]. Upon post-bonding heating, covalent Si-Si bonds can be formed. On the other hand, hydrophilic (SiO_2-SiO_2) bonding can be attributed to the presence of hydroxyl groups in the form of silanol bonds (Si-OH) on the surface [89]. When two hydrophilic surfaces are contacting, the surfaces adhere to each other by hydrogen bonds. If the attached pairs are annealed at higher temperatures, the initial silanol bond will convert to a siloxane network to form a strong bond between the two surfaces (Figure 7b1). Figure 9 shows the hydrophilic bonding process for InP and SOI wafers [67]. When two hydrophobic surfaces are contacting, the surfaces adhere to each other by van der Waals force. The small amount of surface hydroxyl groups assists the Si-Si covalent bond formation after annealing (Figure 7b2) [90]. To increase the amount of hydroxyl groups, additional plasma treatment is necessary [61]. The requirement of ultra-smooth and clean surfaces is one of the disadvantages in the hydrophilic and hydrophobic bonding technologies. Another drawback is the high temperatures needed in the annealing steps. High temperatures may also affect optical coupling between devices by introducing mechanical deformation of the bonded pairs. In hydrophobic bonding, the bonding temperature could be reduced by introducing nano-scale traps for hydrogen atoms on Si surfaces [91]. Such traps could be formed by implanting arsenic ions to the surfaces of Si substrates, treating the Si substrates using diborane or Ar plasma, or sputtering ~1.5 μm amorphous Si layer onto the substrates. These treatments could reduce the bonding temperature from 700 °C to below 450 °C.

As a coupler between on-chip Si waveguides and external optical fibers, grating structures were integrated on a Si substrate using hydrophilic wafer bonding [92]. First, a Si optical mirror was fabricated above the encapsulation layer of the grating coupler due to process convenience. Then the surface of the grating coupler was treated by O_2 plasma, flipped and bonded to an optical chip with waveguides. Thus, the mirror was formed beneath the grating, which enhanced the coupling efficiency up to 69%. Some other applications of hydrophilic wafer bonding include the integration of Si photonics and plasmonic waveguides to improve the coupling efficiency [93], the integration of photoluminescence devices on SOI substrates [94], the encapsulation of fused silica optical gratings [95], and the fabrication of heterojunction photodiodes [96].

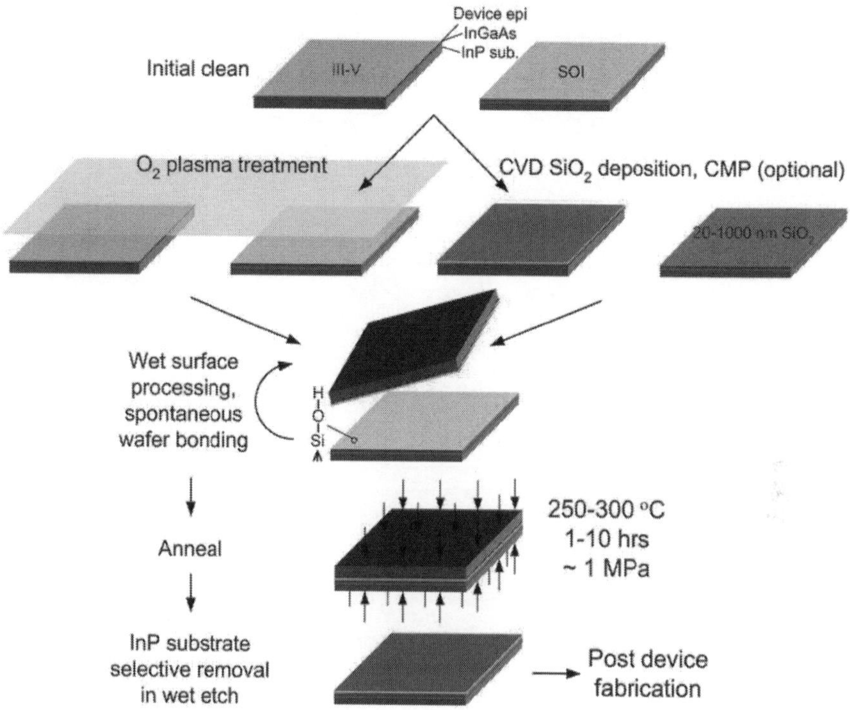

Figure 9. Schematic process flow for hydrophilic bonding of an Si-on-insulator (SOI) wafer and an InP wafer [67].

3.1.3. Surface Activated Bonding

High temperature during or after bonding processes may cause damage at the bonded interface of the optoelectronic components, which was discussed in Section 2.2. One possible solution to address this temperature related issue is to employ the SAB technologies [54,74,97]. The SAB technologies join two clean, smooth and activated surfaces using the adhesion force of surface atoms at room temperature or at low temperatures. The SAB technologies can be categorized into four approaches [14,16], three of which do not need intermediate layers and one approach requires nanoscaled adhesion layer. The three direct bonding approaches will be discussed in this subsection and the fourth one with a nanoadhesion layer is discussed in the next subsection.

For the first direct bonding approach (Figure 10a), the surfaces of bonding pairs are cleaned with an argon (Ar) fast atom beam (FAB), which removes contaminants such as carbon and oxides. Direct adhesion occurs when the surfaces are contacted in an ultra-high vacuum (UHV) at room temperature. High bond strength can be achieved because of the covalent bonding between the atoms of the cleaned and smooth surfaces of the bonded pair [22].

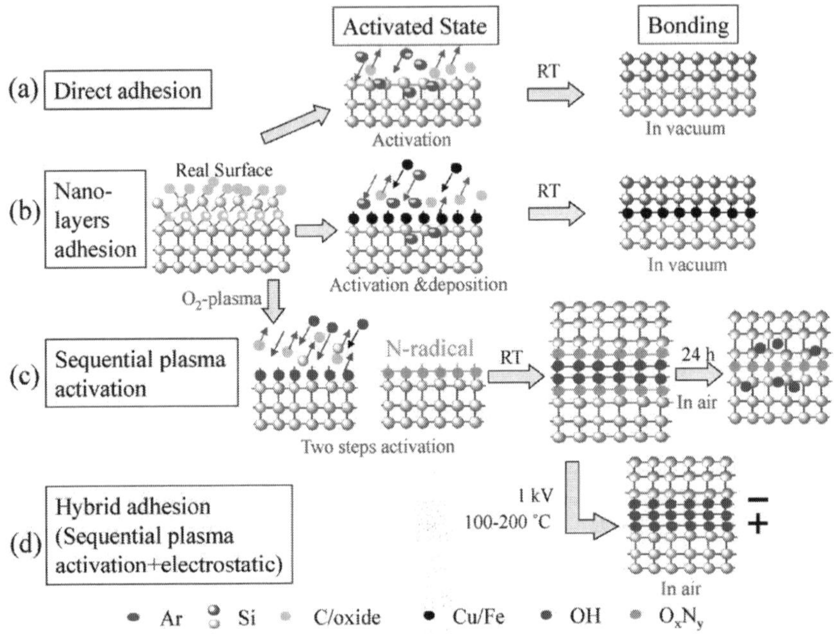

Figure 10. Schematic for surface activated bonding (SAB) technologies that provide atomic-level bonding through: (**a**) Direct adhesion; (**b**) Nano-layers adhesion; (**c**) Sequential plasma activation; (**d**) Hybrid adhesion [16] (Reproduced with permission from Howlader *et al.*, *IEEE J. Sel. Top. Quantum Electron.*; published by IEEE, 2011).

For vertical electrical interconnections, the direct adhesion approach has been demonstrated through bonding of 60 nm height Cu bumps, resulting in high interfacial conductivity [98]. This approach has also been used to bond Cu-filled through-Si vias and Au stud bumps [99]. Using the first SAB approach, material combinations such as Si-InP [20], Si-GaAs [21], Si-GaP, and GaP-GaAs [100] with high bond strength and smooth material interfaces, suitable for optical applications have been reported. There is also demand to bond dissimilar crystalline materials directly for optical isolation in optical systems. A typical example is the bonding of $CeY_2Fe_5O_{12}$ (Ce:YIG) and $LiNbO_3$ [101]. The SAB technique eliminates the difficulties in conventional bonding techniques such as voids formed across the bonded interface [102]. High bond strength has been achieved and the optical isolation performance could be optimized by tuning the Ar

FAB irradiation time. In addition, $LiNbO_3$ has been bonded on Ce:YIG after surface activation by Ar and O_2 mixed plasma, and Ce:YIG has been bonded onto a Si rib waveguide by O_2 plasma activation [103].

In the second direct bonding approach of SAB (Figure 10c), the surfaces are cleaned using radio frequency (RF) RIE and microwave-neutral radicals in low vacuum. Then, the bonding is carried out in a cleanroom atmosphere rather than in UHV. The sequential plasma activation provides highly hydrophilic and reactive surfaces. It also generates nanopores and oxides on surfaces that remove and absorb water from the interface to the bulk material, resulting in strong covalent bonding [13,14,15,16,17,18,19,20,21,22,23,27,28,99,100,102,104,105,106,107].

Semiconducting material pairs such as Si-Si, Si-Ge, and SiO_2-Ge has been bonded using this method. The bonding strength of SAB processed Si-Si pair can be 30 times higher than using hydrophilic bonding [17,25]. To achieve a strong and reliable bond between Ge and Si/SiO_2, heat treatment should be applied after bonding to minimize the hydroxyl groups at the bonded interface [27]. The absorbed water in the bulk of materials across the interface in the sequential plasma activated bonding process results in interfacial voids.

The third direct bonding approach of SAB, named hybrid plasma bonding (Figure 10d), addresses the issue of interfacial voids: the bonded wafers in the second approach are post-treated with an anodic bonding method in a cleanroom atmosphere. This treatment resulted in a void-free bonded interface with high bond strength. The high bond strength is attributed to the adhesion between the hydrophilic and reactive surfaces and the electrostatic forces produced by the anodic step in the bonding process [13,14,15,16,17,18,19,20,21,22,23,27,28,99,100,102,104,105,106,107].

For Si and glass wafers [15], the hybrid plasma bonding provides much stronger adhesion than pure anodic bonding due to the highly hydrophilic, reactive, and smooth surfaces of Si and glass after plasma activation. Also, the temperature requirement of hybrid plasma bonding is lower than anodic bonding. The hybrid plasma bonding has also been demonstrated with other combinations of wafers such Ge, GaAs, SiO_2 and glass [105,108]. The results demonstrate the combined effect of physico-chemically activated surfaces with electrostatic forces to bond ionic materials and Si/glass in the controlled atmosphere of the clean room. The void free and strong heterogeneously integrated interfaces of silicon with other wafers at low temperature can be used in fluidic systems for detecting optical signals from biological species.

3.1.4. Other Direct Bonding Technologies

Cu-Cu has been bonded at room temperature by direct hydrophilic bonding for electrical interfaces [109,110,111]. Chemical mechanical polishing is conducted to provide the surface of Cu with low roughness and good hydrophilicity. Also, plasma-assisted bonding effectively cleans the substrate by removing contaminants and introducing a surface disorder layer. For example, a plasma treated Si substrate possesses a large amount of Si-OH groups, which offer the possibility to increase the bonding strength to over 1.6 J/m^2 through the polymerization of such groups [112,113]. The resulted resistivity at the interconnection was as low as 22.5 $m\Omega \cdot \mu m^2$.

3.2. Indirect Bonding with Intermediate Layers

Although direct wafer bonding in SAB is promising, there are limitations for bonding of certain ionic materials, such as $LiNbO_3$ and $LiTaO_3$ [13,19]. These ionic materials polarize inhomogeneously after surface treatment, which results in repulsive force between the ionic wafers. So the indirect bonding methods are introduced to address this issue.

3.2.1. Adhesive Bonding

In adhesive bonding, often a polymer adhesive is applied between the two wafers to be bonded. Pressure is required to drive the surfaces into intimate contact. Then, the adhesive is cured by either thermal treatment or ultraviolet (UV) light [114], as illustrated before in Figure 7c. The bonding temperature is dependent on the type of adhesive used and can vary from room temperature (for UV cured adhesives) to 100 °C. One of the commonly used adhesive in photonic integration is divinylsiloxane-bis-benzocyclobutene (DVS-BCB), because it has a low curing temperature, high degree of planarization, high optical transmissivity, good thermal stability, excellent chemical stability, and low moisture absorption [115]. The process flow of BCB adhesive bonding is shown in Figure 11. The problems of using an adhesive for wafer bonding are now heighted.

A hermetic seal cannot be created due to the gas permeability of the polymer materials.

The CTEs of adhesives are normally quite different from Si or other semiconductors so that the mechanical reliability cannot be guaranteed.

High conductivity bonding interfaces are challenging to realize (even when using conductive adhesives).

Continuity of optical properties of bonded interfaces is disrupted by adding foreign materials with different optical features (such as different refractive indexes).

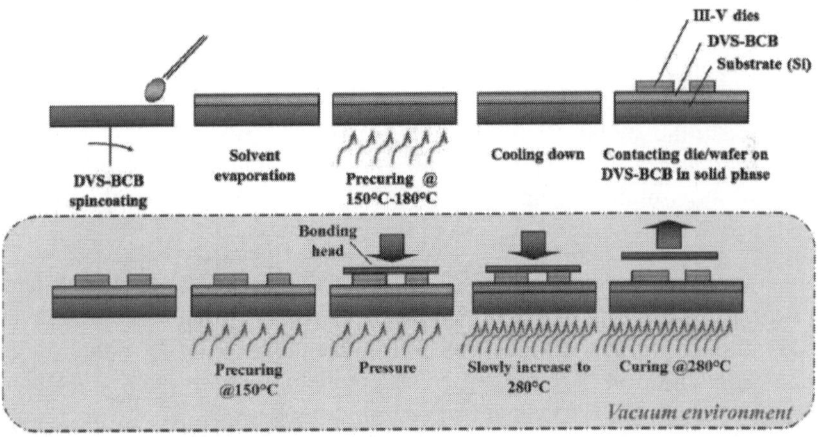

Figure 11. Silicon photonics integration process flow using bis-benzocyclobutene (BCB) bonding [115] (Reproduced with permission from Keyvaninia *et al.*, *Opt. Mater. Express*; published by OSA, 2013).

Since adhesive bonding uses inexpensive polymeric adhesives, it is a cost-effective process and has been implemented in several industrial applications. For instance, miniaturized cameras for cell phones and laptops were fabricated using wafer-level adhesive bonding [116]. Specifically, the adhesive bonding utilized UV-curable epoxies to integrate micro lenses. The bonding process was fast because the epoxy can be cured within 10 s at room temperature. The integrated cameras had a stable optical transmission over 80% for visible light after reliability tests, which was attributed the low outgassing and low shrinkage of the epoxy adhesive. Further, adhesive bonding was used in the fabrication of MEMS micromirror arrays [117], uncooled infrared detectors [117], III-V laser chips with Si waveguides [118,119], integrated photonics chips [120], integrated optical spectrometers and detectors [121,122], and optical circulators [123].

3.2.2. Eutectic Bonding

Eutectic bonding can partially address the low transparency and hermeticity issues of the bonded interfaces in the adhesive bonding by replacing the thick (several hundreds of nanometers to several micrometers) and gas/moisture permeable polymer adhesives. In this method, an intermediate metal layer is used, which can finally change into a eutectic system with the substrate, as shown in Figure 7d. In industry, Au-Si [124] and Al-Ge [125] intermediate layers are mostly used for bonding semiconductor wafers. Sometimes, an Sn-based system is employed for MEMS assembly [126]. Eutectic bonding uses intermediate metal layer that can produce a eutectic system. The eutectic bond is formed by melting a preform (the melting point of the preform is lower than that of its base materials) consisting of a mixture or alloy of two or more dissimilar metals at the bonding interface. However, this approach requires a metal layer between two wafers that may block the optical pathway in assembly of optical devices. Also, an elevated temperature is needed to accelerate the diffusion of the metal.

Eutectic bonding was used to form electrical interconnects in micro-optical systems. For example, in the integrated avalanche photodiode-based detector arrays and cameras [127], In bumps were deposited on the anode contact of each active region. The eutectic bonding between In and Au (on readout circuits) connected the photodiode array to an appropriate readout integrated circuit. The choice of In-Au eutectic system here was due to the low stress in the In bumps at low temperatures (−25 °C to −20 °C), at which the detectors were operated. Also, eutectic bonding was implemented in packaging of GaN LED chips on Si substrates to enhance thermal dissipation [128], in vertical LEDs [129,130], in the development of high-power laser chips [131], as well as in the packaging of thermal imaging sensors [132] and micromirror arrays [133].

3.2.3. Surface Activated Bonding with Nanoadhesion Layers

Since both adhesive bonding and eutectic bonding reduce the transparency of bonded interfaces, the thickness of the intermediate layer should be minimized. The fourth SAB approach using nanometer-thick adhesion layers to bond specific materials such as ionic materials and polymers is now explained (Figure 10b). Bonding of ionic materials is challenging because of their inhomogeneous polarization due to surface activation. In this indirect SAB approach, surface cleaning and deposition of a nanoadhesion layer (for example, a few nanometers of iron (Fe)) are simultaneously performed. The layer controls the surface polarization and enhances the adhesion between the contacting surfaces [13,19]. The two substrates contact with each other spontaneously in low vacuum or UHV and a high bond strength is achieved.

The single crystalline ionic wafers, such as $LiNbO_3$ and $LiTaO_3$, are materials of interest for the fabrication of optoelectronic devices, such as modulators [134]. The ionic materials are required to be integrated with Si substrates. But these materials polarize inhomogeneously after surface activation, which results in repulsive force between the wafers. In order to depolarize the surface, ultrathin (nanometer thickness) Fe films were deposited on the ionic wafers and the activation was prolonged [13,19]. The Fe adhesion layer increases the bonding strength at the bonded interface (8 MPa without Fe nanolayer and 37.5 MPa with it). In addition, the nanometer-thick layer could maintain the optical transparency of the interface, which is very important for optical applications.

To summarize the low-temperature wafer bonding technologies using the SAB technologies, the following are the importance compared with conventional bonding and integration methods.

(1) The ability to bond varieties of solid-state materials with high bond strength.

(2) The applied pressure and process temperature can be lowered. The chemically induced degradation and defect at the bonding interface can be reduced.

(3) The low temperatures used in the process result in high alignment accuracy between bonding pairs.

(4) The realization of mechanical, electrical, and optical connections on the same bonded substrates with minimized optical disruption.

These advantages benefit Si-based micro-optical systems by: (1) incorporating of various types of materials including Si, glass, III-IV materials, ionic materials and polymers; (2) reducing the mismatch of electrical and optical performance at the bonded interface; (3) maximizing the optical coupling efficiency; and (4) enabling small footprint of the fabricated device through wafer-level bonding. However, the long processing time of SAB remains an important disadvantage.

4. FUTURE ASSEMBLY OF OPTICAL SYSTEMS

In the future, aspects in new materials, assembly concepts, and fabrication techniques should be considered in developing micro-optical systems.

First, new materials with unique properties, such as high quantum efficiency, will be integrated with Si to enhance the systems performance such as the increase in signal-to-noise ratio. Second, new integration and assembly concepts will be used to increase the density and reliability of systems. Finally, the cost for development and fabrication of the Si-based systems must be reduced.

4.1. New Materials

Carbon materials have attracted great attention in recent years because of their outstanding mechanical, electrical, thermal, and optical properties. Graphene has properties of strong electron-electron interaction [135] and high photocarrier multiplication [30]. Thus, graphene based photodetectors exhibit ultrafast response. In addition, graphene integrated with micro-/nano-cavities [136,137] or plasmon resonators [138] enhances the detectors' optical absorption and responsivity. Furthermore, hybrid graphene-quantum dot architectures can greatly improve the responsivity of the optical detector. On the other hand, all carbon-based photodetectors have also been demonstrated by integrating zero dimensional graphite quantum dots and 2D graphene sheets [139]. The graphite quantum dots possess high absorptivity, and the graphene sheet has high electrical conductivity. These material properties resulted in a photocurrent responsivity up to 4×10^7 A/W for the all carbon-based photodetector.

CNT, the one dimensional nanomaterial, was found to be sensitive to near infrared (IR) radiation [29]. A CNT based IR detector coated with a pin-hole-free parylene C thin film has been demonstrated with a long lifespan. Also, CNTs has been used as the flip-chip bumps [140,141], and as a filler for TSVs [142].

For the integration of carbon nanomaterials, the CMOS compatible process may be used. There are two possible approaches. In the first approach, low-temperature chemical vapor deposition can be used to directly grow CNTs on Si chips, where electrical and optical components are fabricated. In the second approach, CNTs can be grown on a substrate using high-temperature methods, and then transferred to the target substrate at low temperatures. In terms of electrical contact between carbon nanomaterials and CMOS circuits, CMOS compatible metal (Al) is being studied to replace the materials, such as Au, that are currently used [30].

Another potential new material is MoS_2. A heterojunction photodetector fabricated by combining MoS_2 and amorphous Si (a-Si) has been demonstrated with short response time of ~0.3 ms (Figure 12) [31]. The light is incident from a-Si side, optical absorption occurs in a-Si, and photo-generated electrons diffuse to the underlying MoS_2 layer and are transferred across the MoS_2 layer toward a metal contact. The transient response of this material combination did not show persistent photoconductivity. Thus the device was about 10 times faster than amorphous Si-based ones while its photoresponsivity was 2 to 4 times larger. Such device can also be integrated with MoS_2 transistor based readout circuits, leading to a monolithic technology. Further, WS_2 monolayers are two-dimensional crystals of one atom thickness having direct bandgaps in visible spectrum. At room temperature, the edges of WS_2 monolayers with a zigzag termination exhibit surprisingly strong photoluminescence [32]. This new nanoscaled material could be applied as flexible, transparent, and low-energy optical sources and detectors.

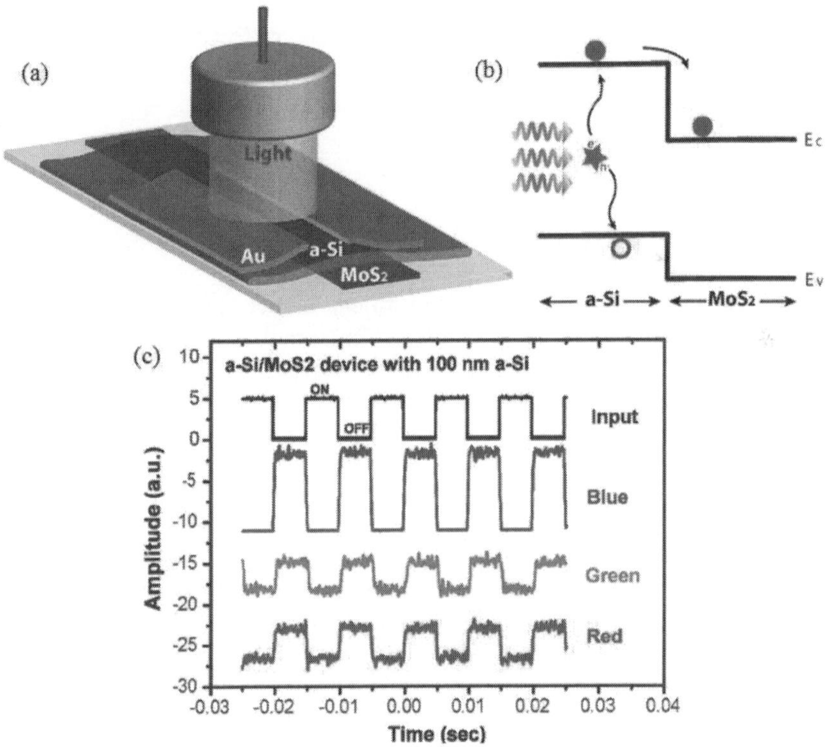

Figure 12. (**a**) Schematic of a-Si/MoS_2 heterojunction photodetector; (**b**) Energy band diagram of a-Si/MoS_2 heterojunction photodetector; (**c**) Transient responses of the device for three incident wavelengths [31] (Reproduced with permission from Esmaeili-Rad *et al.*, *Sci. Rep.*; published by NPG, 2013).

In addition to the abovementioned inorganic materials such as Ge, Si, MoS_2, and WS_2, high performance organic semiconductors with rare earth materials can be used to enhance the responsivity of UV photodetectors [33]. For example, the electron acceptor of tris-(8-hydroxyquinoline) gadolinium and the donor of 1,3,5-tris(3-methylphenyl-phenylamino) triphenyamine (m-MTDATA) has been used in UV detectors.

4.2. Advanced Assembly Concepts

Besides the materials aspect, new assembly concepts are required to improve the performance and reduce the size of Si-based micro-optical systems. Such systems containing optical and electronic devices can be assembled using monolithic or hybrid techniques. In monolithically or hybrid integrated systems, the optical coupling loss can be reduced, the data transmitting rate can be increased (to terabits per second speed) and the power efficiency can be improved (to a few milliwatts per gigabit data [143]). These advantages are attributed to the densely integrated optical and electronic components in a single package [38,144].

4.2.1. Monolithic Integration

Optical components can be monolithically integrated with Si-based CMOS readout circuits. Because all the functional layers in the device are fabricated sequentially on the same substrate, the interconnections between optical and electrical components are fairly short. Thus, high-quality interfaces are easier to realize. Also, the monolithic fabrication of optoelectronic devices is based on current CMOS technology or processing, which reduces the equipment investment.

Ge-on-Si photodetector is an example of a monolithically integrated component (Figure 13) [145]. Here, Si behaves as the waveguide that is interfaced to a Ge optical detector. Such device utilizes the benefits from both Si and Ge: Si's large bandwidth and CMOS compatibility as well as Ge's high responsivity to longer wavelengths than Si. Therefore, the integrated Ge detectors and Si waveguides can provide the detection system with high responsivity and wide bandwidth. This technology requires high-quality epitaxial Ge films grown on Si. One important indication of high-quality Ge on Si is a low threading dislocation density. Several extra steps can be carried out in addition to existing CMOS processes to improve the quality of deposited Ge film on Si. In one approach, a SiGe buffer layer is deposited on the poly-Si gate to reduce the threading dislocation density in the Ge layer which is deposited later.

In another approach, a Ge growth plus annealing method was reported [146]. Such process deposited Ge film more uniformly on Si without the formation of localized Ge islands, which significantly reduces the threading dislocation density.

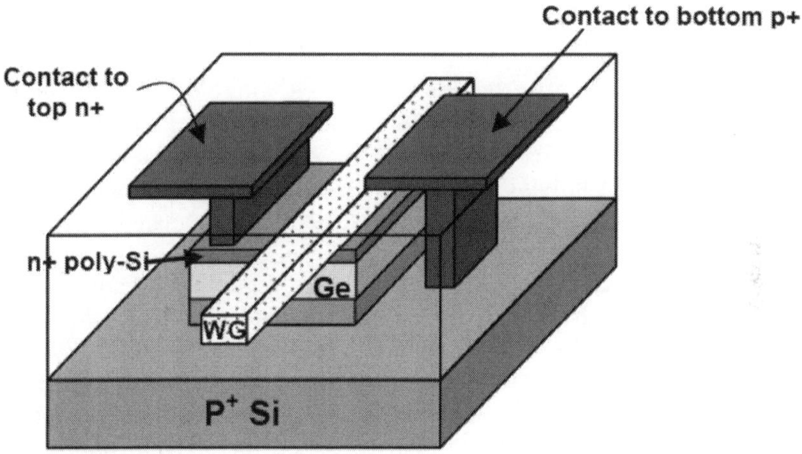

Figure 13. Schematic structure of a waveguide-integrated Ge p-i-n photodetector [145] (Reproduced with permission from Ahn *et al.*, *Opt. Express*; published by OSA, 2007).

Advanced CMOS imagers are three-transistor-based pixel structure which includes Si-based photodectors. They are monolithically integrated by using the existing CMOS processing steps. The electrical interconnection in such a system is small in dimensions and allows the fast conversion of the photogenerated charges to voltages, with signal amplification in the pixels [147,148].

While the standard CMOS processes can be used to fabricate nanometer scale structures and devices, integration of CMOS devices with other optical components on the same chip remains challenging. One of the challenges is the high processing temperature of the optical materials. So the temperature sensitive materials are difficult to be monolithically integrated. Further, MEMS technology consists of many bulk-processing steps such as deep RIE, cavity wet etching, and electroplating. These steps are not compatible with the CMOS technologies. Thus, the process compatibility is the bottleneck in the monolithic integration of MEMS and CMOS components.

4.2.2. Hybrid Integration

Compared to the monolithic integration, hybrid integration of the optical layers with a completed CMOS chip is more convenient as different components can be processed separately and the process compatibility becomes less critical. The fabrication of optical components is independent of the CMOS processes [40,149]. Therefore, the consumption of optical materials could be reduced, resulting in cost reduction.

Optical detectors can effectively be integrated with electronic circuits by embedding the detector in a polymeric waveguide with high coupling efficiency up to ~95% [51,52]. One way to embed waveguide interconnects using polymers is to utilize thin-film optical detectors and bond them to the substrate with electronic components. The polymer waveguide is then fabricated on the top of the bonded substrates and the optical detectors are embedded beneath the waveguide. The advantage of this method is that the waveguide and the embedded optical detector can be integrated onto the substrate with electronic components by post-processing. Hence, the processing steps of the existing electronic boards/chips will not be affected when implementing the optical structures. Finally, the chip tests can be done separately to improve the yield.

To increase the integration density, the optical chip and the Si CMOS chip have to be integrated and assembled in the same module. Using hybrid integration, the yield and performance of the two categories of the components can be independently optimized before integration [72,150]. To realize this hybrid integration, wafer bonding can be used. The wafer bonding processes such as hydrophilic/hydrophobic bonding, adhesive bonding, and SAB are capable of bonding tiny structures made from different materials, resulting in high electrical conductivity and alignment accuracy. Hybrid integration using bonding can make use of the benefits of different materials, such as the optical gain of III-V materials for efficient optical emission [67], the high optical absorption efficiency of Ge for detection [145], and the low loss and high index contrast of Si for lightwave guidance [72,150].

Hybrid-integrated InP/SOI lasers and transmitters based on wafer bonding were reported [151]. Passive rib waveguides, strip waveguides, Bragg reflectors and vertical output couplers were fabricated on the SOI wafer (Figure 14). At the same time, the p-InGaAs contact layer, p-InP cladding layer, InGaAsP quantum wells and InGaAsP separate confinement heterostructure layers were produced on the InP wafer. Afterwards, surface preparation such as polishing was done on the two preprocessed wafers. Hydrophilic bonding was used to join the substrates in a wafer-to-wafer or a chip-to-wafer fashion. Adhesive bonding using BCB polymer was examined as well. Although this approach did not require complicated surface preparation, the yield of devices was 18% lower than using hydrophilic bonding.

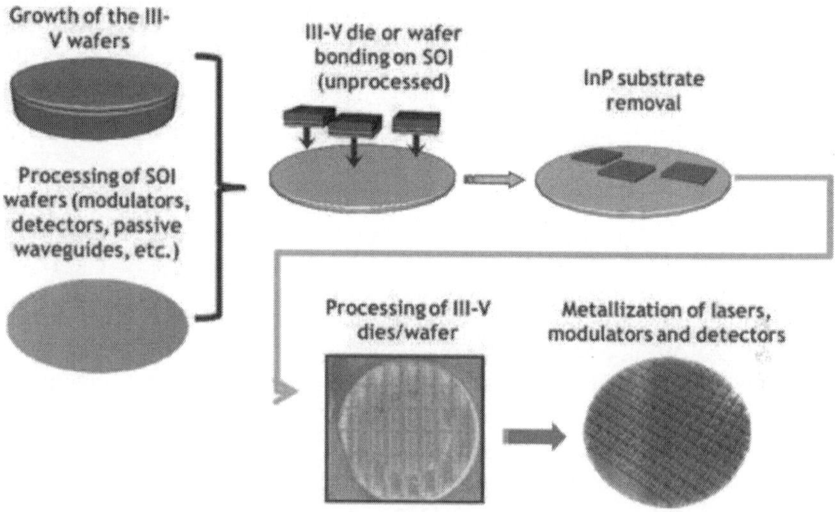

Figure 14. Process flow of hybrid integration of III-V materials on Si [151] (Reproduced with permission from Duan *et al.*, *IEEE J. Sel. Top. Quantum Electron.*; published by IEEE, 2014).

Hybrid integration of thin film devices offers the possibility to fabricate systems with many structures and components. For instance, an integrated lab-on-a-chip system has been demonstrated. This system contained a Si signal processing unit, III-V semiconductor-based optical detection unit, and polymer-based microfluidic unit [152]. The GaAs optical detectors were fabricated on individual substrates and transferred to Si substrates by chip-to-chip bonding. Polymeric materials were used to encapsulate the optical detectors. On top of the polymer encapsulation, the microfluidic components were constructed by soft lithography. The fabrication of individual components used the conventional microelectronic batch processes.

In another example of hybrid integration, cavities were fabricated in the substrates with fabricated photonic devices. The small CMOS chips are placed and bonded inside those cavities to reduce the total size of the products. The photonic substrates with integrated CMOS chips are then bonded to a cap wafer to complete the assembly. The processing simplicity, flexibility, and the ability to integrate wide range of components (microfluidics, optics, electronics, MEMS *etc.*) in hybrid integration make the monolithic integration unappealing [72,150].

Hybrid integration requires technologies such as wafer thinning, vertical interconnection, bumping, molding, dicing and component assembly, and encapsulation. All these technologies need to be optimized and adapted into a modular integrated process flow [153]. Disadvantages of hybrid assembly concept include complex electrical interconnections between the CMOS electronics and optical components. Also, the bonding processes for hybrid integration should be further optimized to improve their compatibility with different types of components to be integrated [40,154].

4.3. Cost Reduction Approaches

The cost of the optical detection systems includes the costs of materials, device fabrication and assembly. The cost may further be increased with the use of new materials such as MoS_2 and their non-conventional integration technologies. Since the CMOS technology is relatively mature and productive, low-cost materials with CMOS process compatibility are required to be developed for the fabrication of optical systems on a Si-based platform in future. For instance, polymers can be used to replace glass for fabricating optical fibers and conductive polymers can replace metals for electrical connections. Polymers may provide robust, continuous and cost-effective chip-to-chip interconnections [155]. Another benefit of polymers is the ease of processing for various devices [156]. A polymer-based hybrid photonic integration platform was developed recently [157]. Such platform mainly used UV-curable epoxy resins and provided flexible optical I/O interfaces coupled with InP active components, film-based optical elements, and on-chip optical fibers. Figure 15 shows a polymer-based optical network unit fabricated on such platform. Another important CMOS-compatible material is SOI. This material platform is used for both optical components (for example, waveguides) and readout electronics. Despite the relatively high cost of SOI wafers, the existing CMOS processing facilities for SOI wafers enable the fabrication to be high-yield and cost-effective [64]. Examples of integrated micro-optical systems have been shown in previous sections.

The fabrication and integration cost is the major part of the total product cost [158]. High volume and efficient production may reduce the product cost. Wafer bonding is one of the large volume and high throughput techniques to achieve low fabrication and integration cost. All dies on a wafer are processed in a series of steps, thereby yielding substantial savings in time, materials, and labor [66,159].

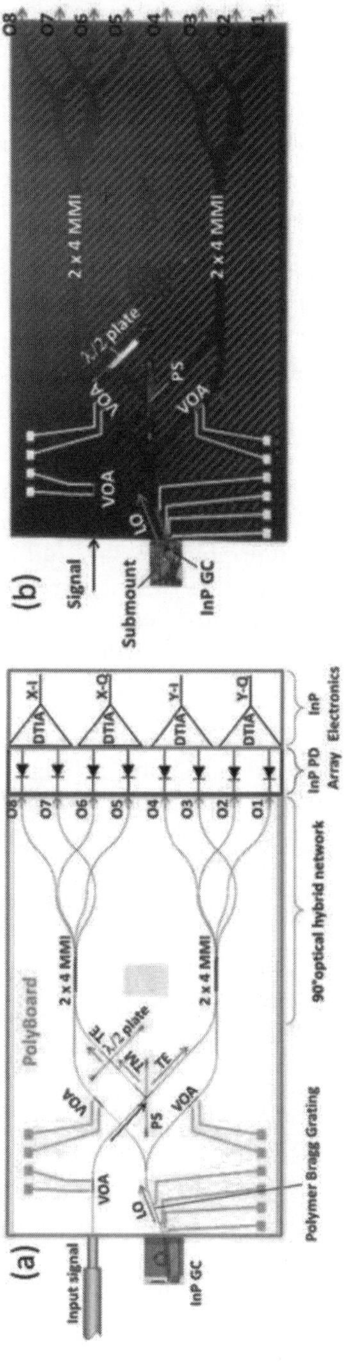

Figure 15. Optical network unit assembled on the polymer-based hybrid integration platform. (**a**) The schematic layout and (**b**) photograph of the assembly [157].

To further reduce the cost of labor and equipment, it is necessary to develop highly automated fabrication and assembly equipment for enhancing manufacturing yield [38]. This equipment also requires to be standardized. For example, when using expensive III-V material, a die-to-wafer bonding process is preferred to reduce the material consumption [64]. The die bonding systems, consisting of the automatic passive optical alignment, materials handling and bonding units, are promising. These criteria enable the processing of many photonic dies in a single run for wafer level hybrid integration [160]. It is also expected that one set of equipment can be used to process different types of substrates/products. Finally, considering the CMOS compatibility, utilizing existing CMOS processing facilities for low cost manufacturing is preferred. Also, when developing new integration equipment, its compatibility with current CMOS lines should be considered.

For both CMOS and photonic dies, low-temperature processing technologies are preferred to preserve material properties and device performance. Low-temperature processes also allow the implementation of materials with low glass transition temperatures (for example, polymers), which may lead to cost reduction [161]. And the time for temperature ramping up and cooling down can be saved to increase the process efficiency.

Finally, the devices should be fabricated using the components with optimized processing steps to meet the application requirements. For instance, the devices should be designed in highly compact form to minimize amount of the optical interconnections and to share common assembly elements such as the heat sink and the substrate [63]. In short distance optical interconnect systems, large diameter fibers could be used to relax the alignment tolerance so that lower the cost of the alignment process [48]. If the products are aimed for low-end market, low cost assembly methods such as liquid encapsulation and transfer molding can be employed.

4.4. Prospected Systems

Recently, a fluorescence spectroscopy based wireless imaging system has been demonstrated using commercially available components for qualitative and quantitative diagnosis of biological cells [162,163,164,165,166]. The system consists of three subsystems: an optical module, an electronic image acquisition module, and an image processing and wireless communication module. The optical module illuminates the biological cells and collects the florescence signal from the cells through optical interfaces between optical components. The image is then acquired by the high-sensitivity image acquisition module, and then sent for further processing via electrical interfaces. The size of the demonstrated system is 2.64 cm × 8.26 cm, as shown in Figure 16. This prototype is too large to fit in gastro-intestinal tract. Therefore miniaturization is required. The miniaturization can be accomplished by shrinking the foot-print of the components, packages and interconnects and by interconnecting them on a single substrate.

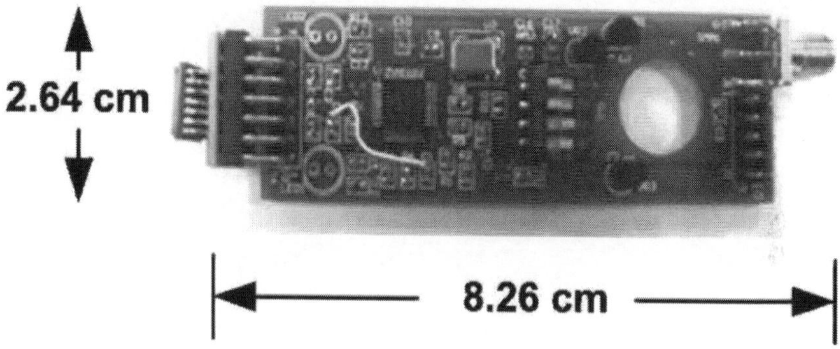

Figure 16. A photograph of the prototype of the imaging acquisition module for diagnosis of biological cells, the imager is on the left mounted perpendicular to the circuit board [162] (Reproduced with permission from Kfouri *et al.*, *IEEE J. Sel. Top. Quantum Electron.*; published by IEEE, 2008.).

Figure 17 shows the schematic diagram of a typical Si-based hybrid integrated wireless health monitoring system with imaging capability of biological cells and proteins. These components guide the light, extract the optical signals, process the signals and transmit the signals wirelessly. The integrated system may provide improved functionality, reliability, performance and reduced cost. Integration and assembly technologies for building up such systems vary a lot depending on their performance requirements, material properties, and application circumstances. Nevertheless, low-temperature wafer bonding is one effective way to realize such system with satisfactory interface qualities.

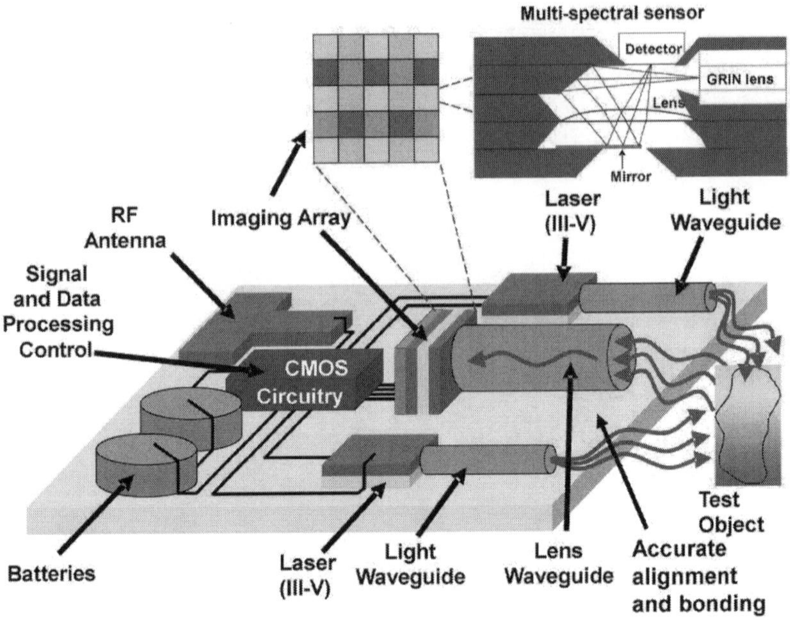

Figure 17. Schematic diagram of an emerging integrated micro-optical system for health monitoring.

5. CONCLUSIONS

Si-based micro-optical systems generate, guide, sense, and image optical signals. In order to shrink the sizes and improve the performance of existing systems, components with small footprints have to be densely integrated. The performance of such systems is determined by the quality of interfaces among electrical and optical components. Highly conductive, low-loss electrical interconnections and optical interfaces with high coupling efficiency are two very important factors. Thus, proper integration and assembly technologies are needed. Electrical interconnections fabricated by wafer bonding offer reduced parasitic electrical parameters. Passive optical alignment of different components is an efficient and cost-effective approach to obtain relatively high coupling efficiency. Conventionally, anodic, hydrophilic and hydrophobic, adhesive and eutectic bonding methods have been used for the integration and assembly Si-based microsystems. However, due to the presence of heat-, pressure- and chemical-sensitive components in systems, these methods are less suitable than the surface activated bonding technologies in terms of interface quality and processing simplicity. The SAB technologies are promising for the electrical and optical interfaces because they do not require high temperature and high external pressure. The SAB technologies enable precise alignment of tiny structures as well as electrically low-loss and optically transparent bonded interfaces. For the future improvement of Si-based micro-optical systems, varieties of new materials such as WS_2 and $MoS_2$with strong photoluminescence or high photo responsivity may be used. Finally, the cost of such systems which is controlled by the materials, processing technologies, and the manufacturing facilities may be reduced by standardization of wafer level processing equipment compatible with CMOS capabilities.

ACKNOWLEDGMENTS

This research is supported by Discovery Grants from the Natural Science and Engineering Research Council of Canada, an infrastructure grant from the Canada Foundation for Innovation, an Ontario Research Fund for Research Excellence Funding Grant, a FedDev of Southern Ontario grant, the Canada Research Chair program, NSERC ResEau strategic network and the NCE IC-IMPACTS. The authors gratefully acknowledge Professor Tadatomo Suga of the University of Tokyo for his ongoing support and contributions to our work.

CONFLICTS OF INTEREST

The authors declare no conflict of interest.

REFERENCES

1 Haes, A.J.; Chang, L.; Klein, W.L.; van Duyne, R.P. Detection of a biomarker for Alzheimer's disease from synthetic and clinical samples using a nanoscale optical biosensor. J. Am. Chem. Soc. 2005, 127, 2264–2271.

2 Zhao, W.; Yao, S.; Hsing, I. A microsystem compatible strategy for viable Escherichia coli detection. Bios. Bioelectron.2006, 21, 1163–1170.

3 Miller, D.A.B. Device requirements for optical interconnects to silicon chips. IEEE Proc. 2009, 97, 1166–1185.

4 Burlacu, D.; Nguyen, L.; Kivilahti, J. Influence of Solder Microstructure and Oxide Layers on High Frequency Electrical Losses of WL-CSP Pb-

Free Interconnections. In Proceedings of IEEE 56th Electronic Components and Technology Conference, San Diego, CA, USA, 30 May–2 June 2006; pp. 577–583.

5 Nemoto, S.; Makimoto, T. Analysis of splice loss in single-mode fibres using a Gaussian field approximation. Opt. Quantum Electron. 1979, 11, 447–457.

6 Yoon, S.W.; Yang, D.W.; Koo, J.H.; Padmanathan, M.; Carson, F. 3D TSV Processes and its Assembly/Packaging Technology. In Proceedings of IEEE International Conference on 3D System Integration, San Francisco, CA, USA, 28–30 September 2009; pp. 1–5.

7 Leib, R.; Topper, M. New Wafer-Level-Packaging Technology Using Silicon-Via-Contacts for Optical and Other Sensor Applications. In Proceedings of IEEE 54th Electronic Components and Technology Conference, Las Vegas, NV, USA, 4 June 2004; pp. 843–847.

8 Liepvre, A.L.; Accard, A.; Poingt, F.; Jany, C.; Lamponi, M.; Make, D.; Lelarge, F.; Fedeli, J.-M.; Messaoudene, S.; Bordel, D. Wavelength selectable hybrid III-V/Si laser fabricated by wafer bonding. IEEE Photonics Technol. Lett. 2013,25, 1582–1585.

9 Barth, P.W. Silicon fusion bonding for fabrication of sensors, actuators and microstructures. Sens. Actuators A Phys.1990, 23, 919–926.

10 Tsau, C.H.; Spearing, S.M.; Schmidt, M.A. Characterization of wafer-level thermocompression bonds. J. Microelectromech. Syst. 2004, 13, 963–971.

11 Dagel, D.J.; Cowan, W.D.; Spahn, O.B.; Grossetete, G.D.; Griñe, A.J.; Shaw, M.J.; Resnick, P.J.; Jokiel, B., Jr. Large-stroke MEMS deformable mirrors for adaptive optics. J. Microelectromech. Syst. 2006, 15, 572–583.

12 Jung, I.W.; Kim, S.; Solgaard, O. High-reflectivity broadband photonic crystal mirror MEMS scanner with low dependence on incident angle and polarization. J. Microelectromech. Syst. 2009, 18, 924–932.

13 Howlader, M.M.R.; Okada, H.; Kim, T.H.; Itoh, T.; Suga, T. Wafer level surface activated bonding tool for MEMS packaging. J. Electrochem. Soc. 2004, 151, G461–G467.

14 Howlader, M.M.R.; Doyle, T.E. Low temperature nanointegration for emerging biomedical applications. Microelectron. Reliab. 2012, 52, 361–374.

15 Howlader, M.M.R.; Kibria, M.G.; Zhang, F.; Kim, M.J. Hybrid plasma bonding for void-free strong bonded interface of silicon/glass at 200 °C. Talanta 2010, 82, 508–515.

16 Howlader, M.M.R.; Selvaganapathy, P.R.; Deen, M.J.; Suga, T. Nanobonding technology toward electronic, fluidic, and photonic systems integration. IEEE J. Sel. Top. Quantum Electron. 2011, 17, 689–703.

17 Howlader, M.M.R.; Suga, T.; Itoh, H.; Lee, T.H.; Kim, M.J. Role of heating on plasma-activated silicon wafers bonding.J. Electrochem. Soc. 2009, 156, H846–H851.

18 Howlader, M.M.R.; Suga, T.; Kim, M.J. Wafer Level and Chip Size Direct Wafer Bonding at Room Temperature. In Proceedings of International Symposia of State-of-the-Art Program on Compound Semiconductors XLI and Nitride and Wide Bandgap Semiconductors for Sensors, Photonics and Electronics V, Honolulu, HI, USA, 4–9 October 2004; pp. 150–160.

19 Howlader, M.M.R.; Suga, T.; Kim, M.J. A novel bonding method for ionic wafers. IEEE Trans. Adv. Packag. 2007, 30, 598–604.

20 Howlader, M.M.R.; Watanabe, T.; Suga, T. Characterization of the bonding strength and interface current of p-Si/n-InP wafers bonded by surface activated bonding method at room temperature. J. Appl. Phys. 2002, 91, 3062–3066.

21 Howlader, M.M.R.; Watanabe, T.; Suga, T. Investigation of the bonding strength and interface current of p-Si/n-GaAs wafers bonded by surface activated bonding at room temperature. J. Vac. Sci. Technol. B 2001, 19, 2114–2118.

22 Howlader, M.M.R.; Zhang, F. Void-free strong bonding of surface activated silicon wafers from room temperature to annealing at 600 °C. Thin Solid Films 2010, 519, 804–808.

23 Howlader, M.M.R.; Zhang, F.; Kibria, M.G. Void nucleation at a sequentially plasma-activated silicon/silicon bonded interface. J. Micromech. Microeng. 2010, 20.

24 Takagi, H.; Maeda, R.; Suga, T. Wafer-scale spontaneous bonding of silicon wafers by argon-beam surface activation at room temperature. Sens. Actuators A Phys. 2003, 105, 98–102.

25 Suga, T.; Kim, T.H.; Howlader, M.M.R. Combined Process for Wafer Direct Bonding by Means of the Surface Activation Method. In Proceedings of IEEE 54th Electronic Components and Technology Conference, Las Vegas, NV, USA, 4 June 2004; pp. 484–490.

26 Takigawa, R.; Higurashi, E.; Suga, T. Passive alignment and mounting of LiNbO3 waveguide chips on Si substrates.IEEE J. Sel. Top. Quantum Electron. 2011, 17, 652–658.

27 Zhang, F.; Kibria, M.G.; Cormier, K.; Howlader, M. Surface and interface characterization of sequentially plasma activated silicon, silicon dioxide and germanium wafers for low temperature bonding applications. ECS Trans. 2010,33, 329–338.

28 Howlader, M.M.R.; Suga, T. Comparative annealing effect on bonded wafers in air and ultrahigh vacuum for microelectromechanical systems/microfluidics packaging. J. Micro Nanolith. MEMS MOEMS 2010, 9.

29 Lai, K.W.C.; Xi, N.; Zhang, J.; Li, G.; Chen, H. Packaging Carbon Nanotube Based Infrared Detector. In Proceedings of IEEE 7th Conference on Nanotechnology, Hong Kong, China, 2–5 August 2007; pp. 778–781.

30 Gan, X.; Shiue, R.-J.; Gao, Y.; Meric, I.; Heinz, T.F.; Shepard, K.; Hone, J.; Assefa, S.; Englund, D. Chip-integrated ultrafast graphene photodetector with high responsivity. Nat. Photonics 2013, 7, 883–887.

31 Esmaeili-Rad, M.R.; Salahuddin, S. High performance molybdenum disulfide amorphous silicon heterojunction photodetector. Sci. Rep. 2013, 3.

32 Gutiérrez, H.R.; Perea-López, N.; Elías, A.L.; Berkdemir, A.; Wang, B.; Lv, R.; López-Urías, F.; Crespi, V.H.; Terrones, H.; Terrones, M. Extraordinary room-temperature photoluminescence in triangular WS2 monolayers. Nano Lett. 2012,13, 3447–3454.

33 Wang, J.B.; Li, W.L.; Chu, B.; Chen, L.L.; Zhang, G.; Su, Z.S.; Chen, Y.R.; Yang, D.F.; Zhu, J.Z.; Wu, S.H.; et al. Visible-blind ultraviolet photo-detector using tris-(8-hydroxyquinoline) rare earth as acceptors and the effects of the bulk and interfacial exciplex emissions on the photo-responsivity. Org. Electron. 2010, 11, 1301–1306.

34 Bernabé, S. Advanced Packaging Concepts for Low-Cost Optoelectronic Devices. In Proceedings of Workshop on Optical Components for Broadband Communication, Stockholm, Sweden, 27 July 2006; p. 635008.

35 Ciftcioglu, B.; Berman, R.; Wang, S.; Hu, J.; Savidis, I.; Jain, M.; Moore, D.; Huang, M.; Friedman, E.G.; Wicks, G. 3-D integrated heterogeneous intra-chip free-space optical interconnect. Opt. Express 2012, 20, 4331–4345.

36 Sainati, R.A.; Moravec, T.J. Estimating high speed circuit interconnect performance. IEEE Trans. Circuits Syst. 1989, 36, 533–541.

37 Tekin, T. Review of packaging of optoelectronic, photonic, and MEMS components. IEEE J. Sel. Top. Quantum Electron.2011, 17, 704–719.

38 Matthews, M.R.; Macdonald, B.M.; Preston, K.P. Optical Components-The New Challenge in Packaging. In Proceedings of IEEE 40th Electronic Components and Technology Conference, Las Vegas, NV, USA, 20–23 May 1990; pp. 206–213.

39 Sancaktar, E.; Bai, L. Electrically conductive epoxy adhesives. Polymers 2011, 3, 427–466.

40 O'Connor, I. Optical Solutions for System-Level Interconnect. In Proceedings of International Workshop on System Level Interconnect Prediction, Paris, France, 14–15 February 2004; pp. 79–88.

41 Tan, Q.; Lee, Y.C. Soldering Technology for Optoelectronic Packaging. In Proceedings of IEEE 46th Electronic Components and Technology Conference, Orlando, FL, USA, 28–31 May 1996; pp. 26–36.

42 RoHS-Restriction of the Use of Certain Hazardous Substances in Electrical and Electronic Equipment Regulations. Available online: http://www.rohs.eu (accessed on 15 August 2015).

43 Tian, Y.; Hutt, D.A.; Liu, C.; Stevens, B. High Density Indium Bmping Through Pulse Plating Used for Pixel X-ray Detectors. In Proceedings of International Conference on Electronic Packaging Technology & High Density Packaging, Beijing, China, 10–13 August 2009; pp. 456–460.

44 Fritzsch, T.; Jordan, R.; Glaw, V.; Topper, M.; Dietrich, L.; Wolf, J.; Ehrmann, O.; Oppermann, H.; Reichl, H.; Wermes, N. Packaging of Radiation and Particle Detectors. In Proceedings of IEEE 56th Electronic Components and Technology Conference, San Diego, CA, USA, 30 May–2 June 2006; pp. 310–314.

45 Garrou, P.; Bower, C.; Ramm, P. Handbook of 3D Integration: Technology and Applications of 3D Integrated Circuits; John Wiley & Sons: Weinheim, Germany, 2008; Vol. 1.

46 Ji, F.; Leppävuori, S.; Luusua, I.; Henttinen, K.; Eränen, S.; Hietanen, I.; Juntunen, M. Fabrication of silicon based through-wafer interconnects for advanced chip scale packaging. Sens. Actuators A Phys. 2008, 142, 405–412.

47 Lapisa, M.; Stemme, G.; Niklaus, F. Wafer-level heterogeneous integration for MOEMS, MEMS, and NEMS. IEEE J. Sel. Top. Quantum Electron. 2011, 17, 629–644.

48 Ho, F.; Hung, V.; Tong, E.; Choi, T.; Chow, A.; Wong, E.; Lui, B.; Yau, K.; Egnisaban, G.; Mangente, T. Ultralow-Cost Optical Transceiver Modules Based on Leadframe Plastic-Package Technology. In Proceedings of IEEE 55th Electronic Components and Technology

Conference, Lake Buena Vista, FL, USA, 31 May–3 June 2005; pp. 1032–1038.

49 Lee, S.H.; Lee, Y.C. Optoelectronic packaging for optical interconnects. Opt. Photonics News 2006, 17, 40–45.

50 Welch, D.F.; Kish, F.A.; Melle, S.; Nagarajan, R.; Kato, M.; Joyner, C.H.; Pleumeekers, J.L.; Schneider, R.; Back, J.; Dentai, A.G. Large-scale InP photonic integrated circuits: Enabling efficient scaling of optical transport networks.IEEE J. Sel. Top. Quantum Electron. 2007, 13, 22–31.

51 Cho, S.-Y.; Seo, S.-W.; Brooke, M.A.; Jokerst, N.M. Integrated detectors for embedded optical interconnections on electrical boards, modules, and integrated circuits. IEEE J. Sel. Top. Quantum Electron. 2002, 8, 1427–1434.

52 Cho, S.-Y.; Hall, J.; Chellappa, A.; Jokerst, N.M.; Brooke, M. High Speed Optical Interconnection Using Embedded PDs on Electrical Boards. In Proceedings of IEEE 56th Electronic Components and Technology Conference, San Diego, CA, USA, 30 May–2 June 2003; pp. 1046–1052.

53 Bhandarkar, S.M. Planar and wafer level packaging of semiconductor lasers and photo detectors for transmitter optical sub-assemblies. U.S. Patent 6,970,491, 29 November 2005.

54 Yamamoto, S.-I.; Higurashi, E.; Suga, T.; Sawada, R. Low-temperature hermetic packaging for microsystems using Au-Au surface-activated bonding at atmospheric pressure. J. Micromech. Microeng. 2012, 22.

55 Higurashi, E.; Chino, D.; Suga, T.; Sawada, R. Au-Au surface-activated bonding and its application to optical microsensors with 3-D structure. IEEE J. Sel. Top. Quantum Electron. 2009, 15, 1500–1505.

56 Higurashi, E. Low-temperature bonding technologies for photonics applications. ECS Trans. 2013, 50, 351–362.

57 Brusberg, L.; Schroder, H.; Topper, M.; Arndt-Staufenbiel, N.; Roder, J.; Lutz, M.; Reichl, H. Thin Glass Based Packaging Technologies for Optoelectronic Modules. In Proceedings of IEEE 59th Electronic Components and Technology Conference, San Diego, CA, USA, 26–29 May 2009; pp. 207–212.

58 Qin, Y.; Turkenburg, D.H.; Barbu, I.; Smaal, W.T.T.; Myny, K.; Lin, W.-Y.; Gelinck, G.H.; Heremans, P.; Liu, J.; Meinders, E.R. Organic thin-film transistors with anodized gate dielectric patterned by self-aligned embossing on flexible substrates. Adv. Funct. Mater. 2012, 22, 1209–1214.

59 Park, J.-S.; Park, K.-B.; Shin, K.-S.; Park, H.-D.; Kim, M.-C.; Kim, J.-R.; Park, S.-J.; Song, Y.-H. Design, fabrication and characterization of an integrated micro ammonia analysis system (IMAAS) with microreactor and in-plane type optical detector based on the Berthelot reaction. Sens. Actuators B Chem. 2006, 117, 516–522.

60 Kopp, C.; Bernabe, S.; Bakir, B.B.; Fedeli, J.; Orobtchouk, R.; Schrank, F.; Porte, H.; Zimmermann, L.; Tekin, T. Silicon photonic circuits: On-CMOS integration, fiber optical coupling, and packaging. IEEE J. Sel. Top. Quantum Electron.2011, 17, 498–509.

61 Dragoi, V.; Mittendorfer, G.; Thanner, C.; Lindner, P. Wafer-level plasma activated bonding: New technology for MEMS fabrication. Microsyst. Technol. 2008, 14, 509–515.

62 Howlader, M.M.R.; Zhang, F.; Deen, M.J.; Suga, T.; Yamauchi, A. Surface activated bonding of copper through silicon vias and gold stud bumps at room temperature. J. Vac. Sci. Technol. A 2011, 29.

63 Roelkens, G.; Liu, L.; Liang, D.; Jones, R.; Fang, A.; Koch, B.; Bowers, J. III-V/silicon photonics for on-chip and intra-chip optical interconnects. Laser Photonics Rev. 2010, 4, 751–779.

64 Roelkens, G.; Van Campenhout, J.; Brouckaert, J.; Van Thourhout, D.; Baets, R.; Romeo, P.R.; Regreny, P.; Kazmierczak, A.; Seassal, C.; Letartre, X. III-V/Si photonics by die-to-wafer bonding. Mater. Today 2007, 10, 36–43.

65 Candler, R.N.; Park, W.-T.; Li, H.; Yama, G.; Partridge, A.; Lutz, M.; Kenny, T.W. Single wafer encapsulation of MEMS devices. IEEE Trans. Adv. Packag. 2003, 26, 227–232.

66 Gooch, R.; Schimert, T. Low-cost wafer-level vacuum packaging for MEMS. MRS Bull. 2003, 28, 55–60.

67 Liang, D.; Roelkens, G.; Baets, R.; Bowers, J.E. Hybrid integrated platforms for silicon photonics. Materials 2010, 3, 1782–1802.

68 Deen, M.J.; Basu, P.K. Silicon Photonics: Fundamentals and Devices; John Wiley & Sons: Chichester, UK, 2012.

69 Beling, A.; Cross, A.S.; Piels, M.; Peters, J.; Zhou, Q.; Bowers, J.E.; Campbell, J.C. InP-based waveguide photodiodes heterogeneously integrated on silicon-on-insulator for photonic microwave generation. Opt. Express 2013, 21, 25901–25906.

70 Park, H.; Sysak, M.N.; Chen, H.-W.; Fang, A.W.; Liang, D.; Liao, L.; Koch, B.R.; Bovington, J.; Tang, Y.; Wong, K. Device and integration technology for silicon photonic transmitters. IEEE J. Sel. Top. Quantum Electron. 2011, 17, 671–688.

71 Le Liepvre, A.; Jany, C.; Accard, A.; Lamponi, M.; Poingt, F.; Make, D.; Lelarge, F.; Fedeli, J.-M.; Messaoudene, S.; Bordel, D. Widely Wavelength Tunable Hybrid III–V/Silicon Laser with 45 nm Tuning Range Fabricated Using a Wafer Bonding Technique. In Proceedings of IEEE 9th International Conference on Group IV Photonics, San Diego, CA, USA, 29–31 August 2012; pp. 54–56.

72 Fang, A.W.; Park, H.; Cohen, O.; Jones, R.; Paniccia, M.J.; Bowers, J.E. Electrically pumped hybrid AlGaInAs-silicon evanescent laser. Opt. Express 2006, 14, 9203–9210.

73 Chen, L.; Dong, P.; Lipson, M. High performance germanium photodetectors integrated on submicron silicon waveguides by low temperature wafer bonding. Opt. Express 2008, 16, 11513–11518.

74 Qin, Y.; Howlader, M.M.R.; Deen, M.J.; Haddara, Y.M.; Selvaganapathy, P.R. Polymer integration for packaging of implantable sensors. Sens. Actuators B Chem. 2014, 202, 758–778.

75 Mehta, A.; Gromova, M.; Rusu, C.; Olivier, R.; Baert, K.; Van Hoof, C.; Witvrouw, A. Novel High Growth Rate Processes for Depositing Poly-SiGe Structural Layers at CMOS Compatible Temperatures. In Proceedings of 17th IEEE International Conference on Micro Electro Mechanical Systems, Maastricht, the Netherlands, 25–29 January 2004; pp. 721–724.

76 Chen, M.; Yuan, L.; Liu, S. Research on low-temperature anodic bonding using induction heating. Sens. Actuators A Phys. 2007, 133, 266–269.

77 Wallis, G.; Pomerantz, D.I. Field assisted glass-metal sealing. J. Appl. Phys. 1969, 40, 3946–3949.

78 Rogers, T. Considerations of anodic bonding for capacitive type silicon/glass sensor fabrication. J. Micromech. Microeng. 1992, 2.

79 Leib, J.; Gyenge, O.; Hansen, U.; Maus, S.; Hauck, K.; Zoschke, K.; Toepper, M. Wafer-Level Glass-Caps for Advanced Optical Applications. In Proceedings of IEEE 61st Electronic Components and Technology Conference, Lake Buena Vista, FL, USA, 31 May–3 June 2011; pp. 1642–1648.

80 Stenchly, V.; Quenzer, H.-J.; Hofmann, U.; Janes, J.; Jensen, B.; Benecke, W. New Fabrication Method of Glass Packages with Inclined Optical Windows for Micromirrors on Wafer Level. Proc. SPIE 2013, 8613.

81 Daschner, R.; Kübler, H.; Löw, R.; Baur, H.; Frühauf, N.; Pfau, T. Triple stack glass-to-glass anodic bonding for optogalvanic

spectroscopy cells with electrical feedthroughs. Appl. Phys. Lett. 2014, 105.

82 Lee, H.J.; Kim, Y.; Yoon, E.-S.; Cho, I.-J. Versatile size and shape microlens arrays with high numerical apertures. J. Microelectromech. Syst. 2014, 23, 771–773.

83 Bargiel, S.; Gorecki, C.; Barański, M.; Passilly, N.; Wiemer, M.; Jia, C.; Froemel, J. 3D Micro-Optical Lens Scanner Made by Multi-Wafer Bonding Technology. Proc. SPIE 2013, 8616.

84 Yoo, S.; Jin, J.-Y.; Ha, J.-G.; Ji, C.-H.; Kim, Y.-K. Two-dimensional optical scanner with monolithically integrated glass microlens. J. Micromech. Microeng. 2014, 24.

85 Surdo, S.; Merlo, S.; Carpignano, F.; Strambini, L.M.; Trono, C.; Giannetti, A.; Baldini, F.; Barillaro, G. Optofluidic microsystems with integrated vertical one-dimensional photonic crystals for chemical analysis. Lab Chip 2012, 12, 4403–4415.

86 Mhaskar, R.; Knappe, S.; Kitching, J. A low-power, high-sensitivity micromachined optical magnetometer. Appl. Phys. Lett. 2012, 101.

87 Kogita, Y.; Hirai, Y.; Tabata, O.; Tsuchiya, T. Double-side-drive electrostatic optical chopper for time-resolved Raman spectroscopy. In Proceedings of International Conference on Optical MEMS and Nanophotonics, Glasgow, Scotland, 17–21 August 2014; pp. 65–66.

88 Tong, Q.-Y.; Schmidt, E.; Gösele, U.; Reiche, M. Hydrophobic silicon wafer bonding. Appl. Phys. Lett. 1994, 64, 625–627.

89 Suni, T.; Henttinen, K.; Suni, I.; Mäkinen, J. Effects of plasma activation on hydrophilic bonding of Si and SiO2. J. Electrochem. Soc. 2002, 149, G348–G351.

90 Tong, Q.-Y.; Schmidt, E.; Gosele, U.; Reiche, M. Hydrophobic silicon wafer bonding. Appl. Phys. Lett. 1994, 64, 625–627.

91 Tong, Q.-Y.; Gan, Q.; Hudson, G.; Fountain, G.; Enquist, P.; Scholz, R.; Gösele, U. Low-temperature hydrophobic silicon wafer bonding. Appl. Phys. Lett. 2003, 83, 4767–4769.

92 Kopp, C.; Augendre, E.; Orobtchouk, R.; Lemonnier, O.; Fedeli, J.-M. Enhanced fiber grating coupler integrated by wafer-to-wafer bonding. J. Lightwave Technol. 2011, 29, 1847–1851.

93 Emboras, A.; Briggs, R.M.; Najar, A.; Nambiar, S.; Delacour, C.; Grosse, P.; Augendre, E.; Fedeli, J.M.; de Salvo, B.; Atwater, H.A. Efficient coupler between silicon photonic and metal-insulator-silicon-metal plasmonic waveguides.Appl. Phys. Lett. 2012, 101.

94 Maruyama, T.; Okumura, T.; Arai, S. Direct wafer bonding of GaInAsP/InP membrane structure on silicon-on-insulator substrate. Jpn. J. Appl. Phys. 2006, 45, 8717–8718.

95 Kalkowski, G.; Zeitner, U.; Benkenstein, T.; Fuchs, J.; Rothhardt, C.; Eberhardt, R. Direct wafer bonding for encapsulation of fused silica optical gratings. Microelectron. Eng. 2012, 97, 177–180.

96 Kanbe, H.; Miyaji, M.; Ito, T. Ge/Si heterojunction photodiodes fabricated by low temperature wafer bonding. Appl. Phys. Express 2008, 1.

97 Higurashi, E.; Imamura, T.; Suga, T.; Sawada, R. Low-temperature bonding of laser diode chips on silicon substrates using plasma activation of Au films. IEEE Photonics Technol. Lett. 2007, 19, 1994–1996.

98 Shigetou, A.; Itoh, T.; Matsuo, M.; Hayasaka, N.; Okumura, K.; Suga, T. Bumpless interconnect through ultrafine Cu electrodes by means of surface-activated bonding (SAB) method. IEEE Trans. Adv. Packag. 2006, 29, 218–226.

99 Howlader, M.M.R.; Yamauchi, A.; Suga, T. Surface activation based nano-bonding and interconnection at room temperature. J. Micromech. Microeng. 2011, 21.

100 Howlader, M.M.R.; Suga, T.; Zhang, F.; Lee, T.H.; Kim, M.J. Interfacial behavior of surface activated p-GaP/n-GaAs bonded wafers at room temperature. Electrochem. Solid State Lett. 2010, 13, H61–H65.

101 Mizumoto, T.; Saito, H.; Sakurai, K.; Shoji, Y. Wafer bonding of magneto-optic garnet and its application to waveguide optical devices. ECS Trans. 2006, 3, 11–22.

102 Howlader, M.M.R.; Suga, T.; Kim, M.J. Room temperature bonding of silicon and lithium niobate. Appl. Phys. Lett.2006, 89.

103 Shoji, Y.; Mizumoto, T.; Yokoi, H.; Hsieh, I.-W.; Osgood, R.M. Magneto-optical isolator with silicon waveguides fabricated by direct bonding. Appl. Phys. Lett. 2008, 92.

104 Howlader, M.M.R.; Iwashita, M.; Nanbu, K.; Saijo, K.; Suga, T. Enhanced Cu/LCP adhesion by pre-sputter cleaning prior to Cu deposition. IEEE Trans. Adv. Packag. 2005, 28, 495–502.

105 Howlader, M.M.R.; Kibria, M.G.; Zhang, F. Hybrid plasma bonding of germanium and glass wafers at low temperature. Mater. Lett. 2010, 64, 1532–1535.

106 Howlader, M.M.R.; Suehara, S.; Takagi, H.; Kim, T.H.; Maeda, R.; Suga, T. Room-temperature microfluidics packaging using sequential plasma activation process. IEEE Trans. Adv. Packag. 2006, 29, 448–456.

107 Howlader, M.M.R.; Suga, T.; Takahashi, A.; Saijo, K.; Ozawa, S.; Nanbu, K. Surface activated bonding of LCP/Cu for electronic packaging. J. Mater. Sci. 2005, 40, 3177–3184.

108 Howlader, M.M.R.; Zhang, F.; Deen, M.J. Formation of gallium arsenide nanostructures in Pyrex glass. Nanotechnology2013, 24.

109 Gueguen, P.; Di Cioccio, L.; Rivoire, M.; Scevola, D.; Zussy, M.; Charvet, A.M.; Bally, L.; Lafond, D.; Clavelier, L. Copper Direct Bonding for 3D Integration. In Proceedings of IEEE International Interconnect Technology Conference, Burlingame, CA, USA, 1–4 June 2008; pp. 61–63.

110 Gueguen, P.; di Cioccio, L.; Gergaud, P.; Rivoire, M.; Scevola, D.; Zussy, M.; Charvet, A.M.; Bally, L.; Lafond, D.; Clavelier, L. Copper direct-bonding characterization and its interests for 3D integration. J. Electrochem. Soc. 2009, 156, H772–H776.

111 Taibi, R.; Di Cioccio, L.; Chappaz, C.; Chapelon, L.-L.; Gueguen, P.; Decham, J.; Fortunier, R.; Clavelier, L. Full Characterization of Cu/Cu Direct Bonding for 3D Integration. In Proceedings of IEEE 60th Electronic Components and Technology Conference, Las Vegas, NV, USA, 1–4 June 2010; pp. 219–225.

112 Moriceau, H.; Bataillou, B.; Morales, C.; Cartier, A.M.; Rieutord, F. Interest of a Short Plasma Treatment to Achieve Si-SiO2-Si Bonded Structures. In Proceedings of the 7th International Symposium on Semiconductor Wafer Bonding Science, Technology, and Applications, Paris, France, 27 April–2 May 2003; pp. 110–117.

113 Moriceau, H.; Rieutord, F.; Morales, C.; Sartori, S.; Charvet, A.M. Surface plasma activation before direct wafer bonding: A short review and recent results. In Proceedings of 207th Meeting of the Electrochemical Society, Quebec, QC, Canada, 16–18 May 2005; pp. 34–49.

114 Niklaus, F.; Stemme, G.; Lu, J.-Q.; Gutmann, R.J. Adhesive wafer bonding. J. Appl. Phys. 2006, 99.

115 Keyvaninia, S.; Muneeb, M.; Stanković, S.; van Veldhoven, P.J.; van Thourhout, D.; Roelkens, G. Ultra-thin DVS-BCB adhesive bonding of III-V wafers, dies and multiple dies to a patterned silicon-on-insulator substrate. Opt. Mater. Express 2013, 3, 35–46.

116 Zoberbier, M.; Hansen, S.; Hennemeyer, M.; Tönnies, D.; Zoberbier, R.; Brehm, M.; Kraft, A.; Eisner, M.; Völkel, R. Wafer Level Cameras-Novel Fabrication and Packaging Technologies. In Proceedings of International Image Sensor Workshop, Bergen, Norway, 26–28 June 2009.

117 Wang, Z. 3-D integration and through-silicon vias in MEMS and microsensors. J. Microelectromech. Syst. 2015, 24, 1211–1244.

118 Snyder, B.W.; O'Brien, P.A. Developments in Packaging and Integration for Silicon Photonics. Proc. SPIE 2013, 8614.

119 Stanković, S.; Jones, R.; Sysak, M.N.; Heck, J.M.; Roelkens, G.; Van Thourhout, D. Hybrid III-V/Si distributed-feedback laser based on adhesive bonding. IEEE Photonics Technol. Lett. 2012, 24, 2155–2158.

120 Lee, J.; Maeda, Y.; Atsumi, Y.; Takino, Y.; Nishiyama, N.; Arai, S. Low-loss GaInAsP wire waveguide on Si substrate with benzocyclobutene adhesive wafer bonding for membrane photonic circuits. Jpn. J. Appl. Phys. 2012, 51.

121 Ryckeboer, E.; Gassenq, A.; Hattasan, N.; Kuyken, B.; Cerutti, L.; Rodriguez, J.-B.; Tournie, E.; Roelkens, G.; Bogaerts, W.; Baets, R. Integrated Spectrometer and Integrated Detectors on Silicon-on-Insulator for Short-Wave Infrared Applications. In Proceedings of Conference on Lasers and Electro-Optics, San Jose, CA, USA, 6–11 May 2012.

122 Roelkens, G.; Green, W.M.J.; Kuyken, B.; Liu, X.; Hattasan, N.; Gassenq, A.; Cerutti, L.; Rodriguez, J.B.; Osgood, R.M.; Tournié, E. III–V/silicon photonics for short-wave infrared spectroscopy. IEEE J. Quantum Electron. 2012, 48, 292–298.

123 Ghosh, S.; Keyvaninia, S.; Van Roy, W.; Mizumoto, T.; Roelkens, G.; Baets, R. Adhesively bonded Ce:YIG/SOI integrated optical circulator. Opt. Lett. 2013, 38, 965–967.

124 Wolffenbuttel, R.F. Low-temperature intermediate Au-Si wafer bonding; eutectic or silicide bond. Sens. Actuators A Phys. 1997, 62, 680–686.

125 Sood, S.; Farrens, S.; Pinker, R.; Xie, J.; Cataby, W. Al-Ge eutectic wafer bonding and bond characterization for CMOS compatible wafer packaging. ECS Trans. 2010, 33, 93–101.

126 Liu, H.; Salomonsen, G.; Wang, K.; Aasmundtveit, K.E.; Hoivik, N. Wafer-level Cu/Sn to Cu/Sn SLID-bonded interconnects with increased strength. IEEE Trans. Compon. Packag. Manuf. Technol. 2011, 1, 1350–1358.

127 Itzler, M.A.; Entwistle, M.; Krishnamachari, U.; Owens, M.; Jiang, X.; Slomkowski, K.; Rangwala, S. SWIR Geiger-Mode APD Detectors and Cameras for 3D Imaging. Proc. SPIE 2014, 9114.

128 Wen, S.-Y.; Hu, H.-L.; Tsai, Y.-J.; Hsu, C.-P.; Lin, R.-C.; Horng, R.H. A novel integrated structure of thin film GaN LED with ultra-low thermal resistance. Opt. Express 2014, 22, A601–A606.

129 Choi, E.; Kim, A.; Cui, Y.; Chae, S.J.; Nam, M.; Kwon, S.H.; Cha, Y.W.; Pyo, S.G. Eutectic bonding utilizing radio frequency induction heating for fabricating vertical light-emitting diodes. J. Nanosci. Nanotechnol. 2015, 15, 8533–8536.

130 Hao, G.-D.; Wang, X.-L. Enhancement of light-extraction efficiency in AlGaInP light-emitting diodes using evanescent wave coupling effect. Appl. Phys. Lett. 2013, 103.

131 McDougall, S.; McKee, A.; Eddie, I.; Tweedie, A.; Meredith, W. Development of High Power Laser Technology: 915 nm Mini-Bars for Fibre Laser Pumping and Red Lser Bars for Cinema/Projector Applications. In Proceedings of High Power Diode Lasers and Systems Conference, Coventry, UK, 16–17 October 2013; pp. 22–23.

132 Han, M.-S.; Han, S.M.; Kim, H.J.; Shin, J.C.; Ahn, M.S.; Kim, H.W.; Han, Y.H. Low-Cost Compact Thermal Imaging Sensors for Body Temperature Measurement. Proc. SPIE 2013, 8704.

133 Lani, S.; Bayat, D.; Pétremand, Y. 2D Electrostatic Micromirror Array with High Field Factor for High-Power Application. Proc. SPIE 2013, 8616.

134 Griffel, G.; Ruschin, S.; Croitoru, N. Linear Electrooptic Effect in Sputtered Polycrystalline LiNbO3 Films. In Proceedings of International Congress on Optical Science and Engineering, Hamburg, Germany, 19–23 September 1988; pp. 228–232.

135 Kotov, V.N.; Uchoa, B.; Pereira, V.M.; Guinea, F.; Neto, A.H.C. Electron-electron interactions in graphene: Current status and perspectives. Rev. Mod. Phys. 2012, 84.

136 Gan, X.; Mak, K.F.; Gao, Y.; You, Y.; Hatami, F.; Hone, J.; Heinz, T.F.; Englund, D. Strong enhancement of light-matter interaction in graphene coupled to a photonic crystal nanocavity. Nano Lett. 2012, 12, 5626–5631.

137 Furchi, M.; Urich, A.; Pospischil, A.; Lilley, G.; Unterrainer, K.; Detz, H.; Klang, P.; Andrews, A.M.; Schrenk, W.; Strasser, G. Microcavity-integrated graphene photodetector. Nano Lett. 2012, 12, 2773–2777.

138 Liu, Y.; Cheng, R.; Liao, L.; Zhou, H.; Bai, J.; Liu, G.; Liu, L.; Huang, Y.; Duan, X. Plasmon resonance enhanced multicolour photodetection by graphene. Nat. Commun. 2011, 2.

139 Cheng, S.-H.; Weng, T.-M.; Lu, M.-L.; Tan, W.-C.; Chen, J.-Y.; Chen, Y.-F. All carbon-based photodetectors: An eminent integration of graphite quantum dots and two dimensional graphene. Sci. Rep. 2013, 3.

140 Fu, Y.; Qin, Y.; Wang, T.; Chen, S.; Liu, J. Ultrafast transfer of metal-enhanced carbon nanotubes at low temperature for large-scale electronics assembly. Adv. Mater. 2010, 22, 5039–5042.

141 Wang, T.; Jiang, D.; Chen, S.; Jeppson, K.; Ye, L.; Liu, J. Formation of three-dimensional carbon nanotube structures by controllable vapor densification. Mater. Lett. 2012, 78, 184–187.

142 Wang, T.; Chen, S.; Jiang, D.; Fu, Y.; Jeppson, K.; Ye, L.; Liu, J. Through-silicon vias filled with densified and transferred carbon nanotube forests. IEEE Electron Device Lett. 2012, 33, 420–422.

143 Reichl, H.; Aschenbrenner, R.; Potter, H. Hetero system integration-Enabling technology for future products. German Technology-Electronics Production Equipment; VDMA Verlag: Frankfurt/Main, Germany, 2008; pp. 45–49.

144 Heck, M.; Bauters, J.; Davenport, M.; Doylend, J.; Jain, S.; Kurczveil, G.; Srinivasan, S.; Tang, Y.; Bowers, J. Hybrid silicon photonic integrated circuit technology. IEEE J. Sel. Top. Quantum Electron. 2013, 4, 1–18.

145 Ahn, D.; Hong, C.-Y.; Liu, J.; Giziewicz, W.; Beals, M.; Kimerling, L.C.; Michel, J.; Chen, J.; Kńrtner, F.X. High performance, waveguide integrated Ge photodetectors. Opt. Express 2007, 15, 3916–3921.

146 Michel, J.; Liu, J.; Kimerling, L.C. High-performance Ge-on-Si photodetectors. Nat. Photonics 2010, 4, 527–534.

147 Rogalski, A. Progress in focal plane array technologies. Prog. Quantum Electron. 2012, 36, 342–473.

148 Rogalski, A. History of infrared detectors. Opto Electron. Rev. 2012, 20, 279–308.

149 El-Desouki, M.; Jamal Deen, M.; Fang, Q.; Liu, L.; Tse, F.; Armstrong, D. CMOS image sensors for high speed applications. Sensors 2009, 9, 430–444.

150 Liang, D.; Bowers, J.E. Photonic integration: Si or InP substrates? Electron. Lett. 2009, 45, 578–581.

151 Duan, G.-H.; Jany, C.; le Liepvre, A.; Accard, A.; Lamponi, M.; Make, D.; Kaspar, P.; Levaufre, G.; Girard, N.; Lelarge, F. Hybrid III-V on silicon lasers for photonic integrated circuits on silicon. IEEE J. Sel. Top. Quantum Electron. 2014, 20, 158–170.

152 Song, F.; Xiao, J.; Udawala, F.; Seo, S.-W. Heterogenous integration of a thin-film GaAs photodetector and a microfluidic device on a silicon substrate. J. Micromech. Microeng. 2011, 21.

153 Bock, K.; Wolf, J.; Reichl, H. Trends in RF & Wireless Packaging. In Proceedings of the International Conference on Compound Semiconductor Manufacturing Technology (Sharing Ideas Throughout the Industry), Miami, FL, USA, 3–6 May 2004; p. 10A.

154 Ma, C.L.F.; Deen, M.J.; Tarof, L.E.; Yu, J.C.H. Temperature dependence of breakdown voltages in separate absorption, grading, charge, and multiplication InP/InGaAs avalanche photodiodes. IEEE Trans. Electron Devices 1995, 42, 810–818.

155 Cho, I.-K.; Yun, J.-H.; Jeong, M.-Y.; Park, H.-H. Optical chip-to-chip link system by using optical wiring method for reducing EMI. IEEE Trans. Adv. Packag. 2010, 33, 722–728.

156 Liu, F.; Guidotti, D.; Sundaram, V.; Mahajan, S.; Huang, Z.; Chang, Y.-J.; Chang, G.K.; Tummala, R.R. Material and Process Challenges in Embedding Polymeric Waveguides and Detectors in System on Package (SOP). In Proceedings of IEEE 9th International Symposium on Advanced Packaging Materials: Processes, Properties and Interfaces, Atlanta, GA, USA, 24–26 March 2004; pp. 89–94.

157 Zhang, Z.; Felipe, D.; Katopodis, V.; Groumas, P.; Kouloumentas, C.; Avramopoulos, H.; Dupuy, J.-Y.; Konczykowska, A.; Dede, A.; Beretta, A. Hybrid photonic integration on a polymer platform. Photonics 2015, 2, 1005–1026.

158 Qin, Y.; Kwon, H.-J.; Howlader, M.M.R.; Deen, M.J. Microfabricated electrochemical pH and free chlorine sensors for water quality monitoring: Recent advances and research challenges. RSC Adv. 2015, 5, 69086–69109.

159 Deen, M.J.; Pascal, F. Electrical characterization of semiconductor materials and devices-Review. J. Mater. Sci. Mater. Electron. 2006, 17, 549–575.

160 Boudreau, R.; Zhou, P.; Bowen, T. Wafer scale photonic-die attachment. IEEE Trans. Compon. Packagi. Manuf. Technol.1998, 21, 136–139.

161 Qin, Y.; Alam, A.U.; Pan, S.; Howlader, M.M.R.; Ghosh, R.; Selvaganapathy, P.R.; Wu, Y.; Deen, M.J. Low-temperature solution processing of palladium/palladium oxide films and their pH sensing performance. Talanta 2015, 146, 517–524.

162 Kfouri, M.; Marinov, O.; Quevedo, P.; Faramarzpour, N.; Shirani, S.; Liu, L.W.-C.; Fang, Q.; Deen, M.J. Toward a miniaturized wireless fluorescence-based diagnostic imaging system. IEEE J. Sel. Top. Quantum Electron. 2008, 14, 226–234.

163 Cheng, Z.; Deen, M.J.; Peng, H. A low-power gateable vernier ring oscillator time-to-digital converter for biomedical imaging applications. IEEE Trans. Biomed. Circuits Syst. 2015, PP, 1–10.

164 Cheng, Z.; Zheng, X.; Deen, M.J.; Peng, H. Development of a high performance digital silicon photomultiplier (dSiPM) for ToF PET imaging. J. Nucl. Med. 2015, 56, 603.

165 Zheng, X.; Cheng, Z.; Deen, M.J.; Peng, H. Improving the spatial resolution in CZT detectors using charge sharing effect and transient signal analysis: Simulation study. Nucl. Instrum. Meth. A 2015, 808, 60–70.

166 Zheng, X.; Cheng, Z.; Deen, M.J.; Peng, H. Investigation of the sub-pixel spatial resolution and charge-sharing effect in CZT detectors for PET imaging. J. Nucl. Med. 2015, 56, 1873.

CHAPTER 8

Silicon Photonic Crystals Towards Optical Integration

Zhi-Yuan Li[1], Chen Wang[1] and Lin Gan[1]

[1] Laboratory of Optical Physics, Institute of Physics, Chinese Academy of Sciences, Beijing, China

1. INTRODUCTION

During the past two decades, there have been great interests in developing ways to manipulate photons at nanoscale, realizing optical integrations, developing smaller, faster, and more efficient optoelectronic devices for the purpose of next-generation optoelectronic technology. Great progresses have been made in exploring photonic crystals (PCs) [1,2], plasmonic structures [3,4], and other nanophotonic devices for applications. However, plasmonic structures always involve some metal cells and are subject to strong energy dissipation and absorption loss in optical frequencies. Since silicon has a large refraction index and low loss in the infrared wavelength, it becomes an important optical material that has been widely used for integrated photonics applications. Meanwhile, silicon dominates microelectronics and this makes the silicon-based optical devices have the advantage to integrate with electronic devices.

Among all the semiconductor-based optical devices, a class of integrated optical devices that are built in the platform of periodically patterned silicon structures (namely, silicon PCs) are now attracting much attention [5–7]. Analogous to real crystal, electromagnetic (EM) wave is strongly modulated in PC by means of periodic Bragg scattering. Photonic band gaps (PBGs), which can prevent light from propagation in certain direction for a certain range of wavelengths, are formed similarly to electron band gaps. If we introduce a line defect or a point defect in PC, a defect state will take place within the PBG, where light is strongly localized around the defect. These defects can serve as a high efficient waveguide channel or as a micro-cavity with a high-quality (high-Q) factor. The mismatch of the PBGs spatial inversion symmetry breaking and could lead to the optical isolation in any device where the forward and backward transmissivity of light is very much different.In addition, the transmission bands also provide remarkable dispersion properties due to strong Bragg scattering, and negative refraction, self-collimation, superprism and many other anomalous transport behaviors [8–13] can be achieved by engineering the unit cell geometry of PCs. In this review, we briefly introduce the theoretical background of the light propagation in PC and show our recent results on design, fabrication, and characterization of several basic integrated optical devices in the platform of infrared silicon PC slab.

2. THEORETICAL AND NUMERICAL TOOLS

In this section, we briefly introduce the theoretical background and the numerical methods for our study of PC. The propagation of electromagnetic waves in PC is governed by the Maxwell equations [14]. For the sake of simplicity, we only consider a nonmagnetic linear system. In particular, the dielectric constant εε is independent of frequency and we neglect any absorption of electromagnetic waves by the material. Furthermore, there are no free charges or currents in our system. With all of the assumptions, the magnetic field within the PC satisfies the following equation that directly originates from the Maxwell equations [15]:

$$\nabla \times \left\{ \frac{1}{\varepsilon(\mathbf{r})} \nabla \times \mathbf{H}(\mathbf{r}) \right\} = \frac{\omega^2}{c^2} \mathbf{H}(\mathbf{r}). \tag{1}$$

Since our system involves a periodic dielectric function εε, we can apply Bloch-Floquet theorem to our situation, which means that the solutions can be expressed as:

$$\boldsymbol{H}_k(\boldsymbol{r}) = \boldsymbol{u}_k(\boldsymbol{r}) \mathrm{e}^{\mathrm{i}(\mathbf{k}\cdot\mathbf{r}-\omega t)}, \tag{2}$$

where k is the Bloch wave vector and uk(r) is a periodic function of position. This type of solutions are periodic as a function of k.

Several theoretical methods have been developed to handle different problems for PC structures, such as the plane-wave expansion method (PWEM) [15–18], transfer matrix method (TMM) [19], finite-difference time-domain (FDTD) method [20,21], and multiple scattering method [22,23]. Each method exhibits its own benefits and drawbacks. The PWEM is the earliest method applied to PC [15], and has shown its great power in the discovery of three-dimensional (3D) diamond-lattice PCs that have a complete PBG [15]. However, this method has a severe limitation in that it can only deal with the photonic band structures. The TMM is an efficient approach that was designed particularly to calculate the transmission spectra of PC, but it can also be used to solve the photonic band structures. This approach works based on the finite-difference scheme in the real space [19]. Later on, a plane-wave based transfer-matrix method (PWTMM) was developed by Li et al. [24–26]. This method works on the plane-wave space and uses plane wave functions (representing Bragg waves) to describe both the EM fields and dielectric functions. This approach can handle a broad range of general PC problems. In addition to the regular solutions of photonic band structures and transmission/reflection/absorption spectra, this approach can efficiently solve the Bloch wave scattering at the interface between semi-infinite PC structures, because it can also work in the Bloch mode space [27]. So far, other methods have not developed such a peculiar capability. For this reason, it can calculate efficiently modal coupling with multimode PC waveguides [28], transmission efficiency through general two-dimensional (2D) and 3D waveguide bends [29], and band diagrams and field profiles of PC surface states [30]. On the other hand, this method has adopted advanced numerical and mathematical analysis tools to enhance numerical convergence and accuracy, and it has shown its superior power in dealing with some metal PC structures compared with other methods [31]. Recently, this method was also extended to solve nonlinear optical problems in ferroelectric PC structures [32,33]. The FDTD method is a very popular and universal approach in numerical simulations of various PC problems. In addition to the regular band structures and optical spectra calculations, this technique can govern the EM field evolution with time in arbitrary PC structures with infinite or finite structural domain. The reason is that the technique works in the time domain and directly solves the Maxwell equations.

Many free software packages and commercial software packages have been developed worldwide and they are widely used in numerical simulations and solutions of different PC problems. Our group has also developed homemade codes based on several methods including the PWEM, PWTMM, FDTD, and multiple scattering method. In addition, we also utilize publicly available free software packages as they are more numerically economic or have better numerical efficiency. In our case, we use MIT Photonic-Bands (MPB) package [16] to compute the photonic band structures and use MEEP, a free FDTD simulation software package developed at MIT [34] to calculate transmission spectra and model electromagnetic wave transport features in the 2D PC structures and devices.

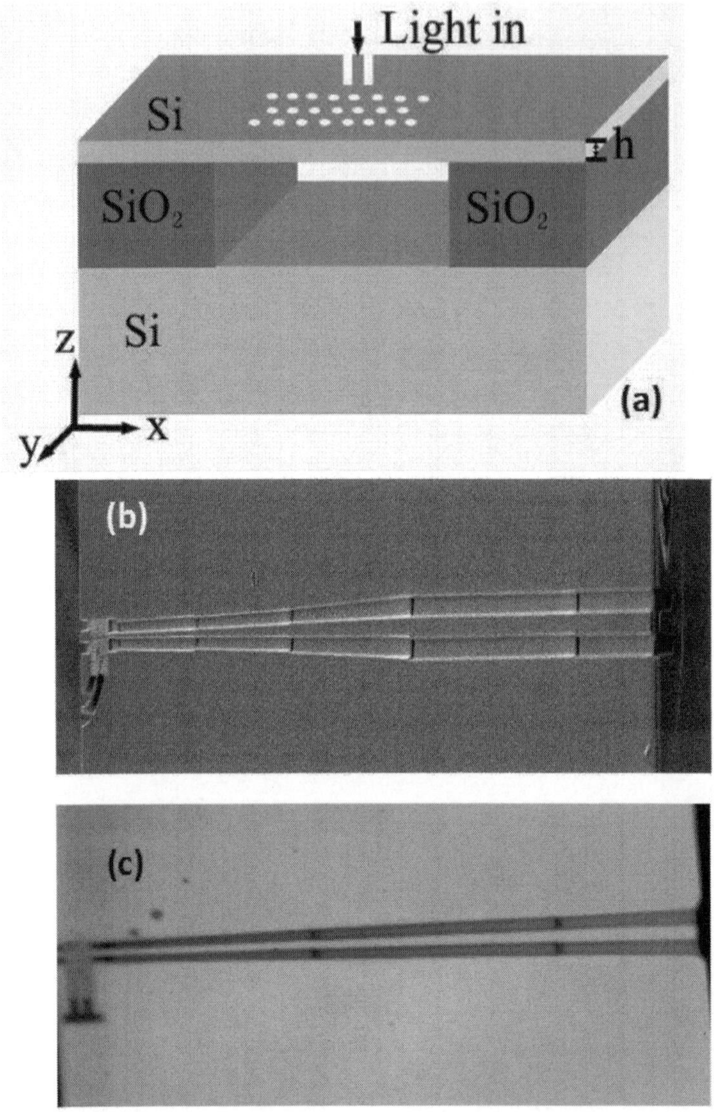

Figure 1. a) Schematic view of a 2D air-bridged PC structures with an input silicon waveguide. The whole structures are fabricated in SOI wafer. The air-bridged structures are formed by HF wet etching; (b) and (c) are the top-view SEM and optical microscopy image of a practical PC sample used in experiment. The long adiabatically tapering ridge waveguide connected with the PC structure can be clearly visualized.

3. SAMPLE FABRICATION AND OPTICAL CHARACTERIZATION

After the discovery of PC, many novel devices have been proposed to control light and implement specific functionality of information processing. 2D air-bridged silicon PC slab [Fig. 1(a)] is an excellent platform to fabricate PC integrated optical devices. This system involves a silicon membrane suspended in air, which confines light by high index contrast in the vertical direction, while the periodic structures in the slab give a strong in-plane confinement of light through PBGs. In most cases, a typical PC structure is a kind of periodic array of air holes etched in a silicon-on-insulator (SOI) wafer by microfabrication techniques. The SOI wafer has a $Si/SiO_2/Si$ structure. In our case, it has a 220 nm thick silicon top layer and a 3 μm buried silica layer on top of a 0.5 mm thick single crystal silicon wafer. We directly use focused ion beam (FIB) lithography to drill air holes in the silicon membrane or use electron-beam lithography (EBL) to define PC patterns in a thin film of polymethylmethacrylate (PMMA), and then transfer the patterns into the silicon membrane by inductively coupled plasma (ICP) etching under the atmosphere of SF_6 and C_4F_8 gases. Figures 1(b) and (c) are the top-view scanning electron microscope (SEM) picture of our PC structures fabricated by FIB. By utilizing state-of-the-art microfabrication techniques, the optical properties of the periodic array of air holes can be easily and accurately controlled. For instance, one can change the diameters of certain holes or omit to etch one or several holes at certain places. These procedures can allow for engineering of the linear and point defect characteristics. After the air holes PC structures get done, we use HF acid wet etching to remove the buried oxide layer under the silicon membrane PC structures to form air-bridged structures. Usually, wide silicon wire waveguides (also with the air-bridged geometry) close to the interface of PC structures are used as the input and output infrared light beam channels, as shown in Fig. 1. These wire waveguides are further connected with long adiabatically tapering ridge silicon waveguides (each about 0.2 mm long) to allow easy coupling with external infrared signals from single mode optical fibers. As a result, a typical PC sample has a total length of about 0.5 mm and the input and output ends are carefully polished to enhance the coupling efficiency of input and output infrared signals.

The transmission spectra of a PC structure effectively reflect its optical properties. To get this important physical quantity, we have set up an experimental apparatus that involves several functional components. The overall measurement setup is schematically illustrated in Fig. 2(a), while a picture of the corresponding real system is displayed in Fig. 2(b). As shown in Fig. 2(a), the PC samples are placed in the center of the stage, with its two sides connected with the input and output optical fibers. The input optical signal comes from a continuous wave tunable semiconductor laser with the wavelength ranging from 1500 to 1640 nm, launched into one facet of the ridge waveguide via a single-mode lensed fiber. Power meter is used to detect the optical signals transmitted through the PC structures and emitted from the output side. The measurement is made with TE polarization (electric field parallel to the slab plane) since it has a complete band gap in silicon PC slabs. The measurement data are normalized by a ridge waveguide on silicon with the same length and width to yield the final transmission spectra for a specific PC structure.

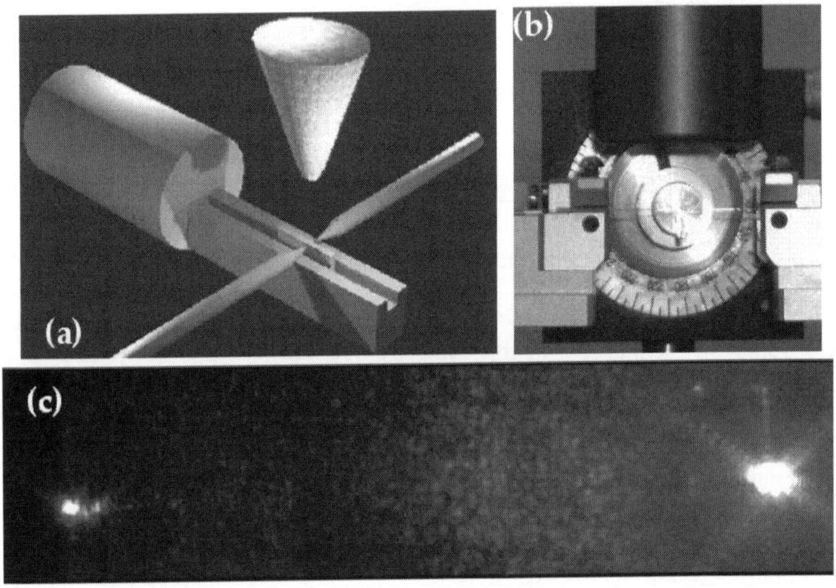

Figure 2. a) Schematic view and (b) experimental setup for the optical characterization of infrared 2D silicon PC slab structures; (c) typical optical microscopy picture recorded by the CCD camera for the PC sample as displayed in Fig. 1(b).

In addition to the measurement of transmission spectra, our experimental setup can offer another big power: it allows for easy and convenient direct monitoring of the transport path of infrared light through PC devices. As depicted in Fig. 2(a), a charge-coupled device (CCD) camera is mounted above the sample and it can *in situ* monitors the transport property by imaging the roughness induced scattering infrared light from the surface of the PC structures. The long-focus microscope objective connected with the CCD camera is shown in the upper part of Fig. 2(b). The ray trace can be directly visualized by the camera to yield images at a personal computer monitor, and this gives the researcher a rough but direct estimate abouthow much the infrared signal has gone into the PC structures. The idea is simple: if the infrared light is coupled into the PC sample with a sufficiently high efficiency, infrared light can transport along the input ridge waveguide, PC devices, and the output ridge waveguide. Significant scattering of infrared signal off the sample can take place and is collected by the CCD camera and visualized *in situ* by the monitor. The strongest scattering occurs at the discontinuity interface, including the end facets of input and output ridge waveguides and the connection section between ridge waveguides and PC structures. Even within the PC structures, remarkable scattering still takes place because of the inevitable roughness on the surface of silicon slab and within the inner walls of air holes. On the other hand, if the infrared signal coupled into the PC samples is weak, then the overall scattering light that can be collected by the CCD camera is also very weak, and no picture can be visualized in the monitor. Figure 2(c) shows a typical microscopy picture recorded by the CCD camera. Such an experimental setup is very convenient to adjust the precise position of the input and output optical fibers, so that they have a precise point-to-point alignment with about 250 nm thick PC samples to allow for high-efficiency optical coupling. The infrared signal transporting through the PC samples usually has a power level of micro-walt when the input signal from semiconductor laser is on the power level of milli-walt. The signal has been already sufficiently strong to allow for transmission spectrum measurement and CCD camera monitoring with high signal-to-noise ratio.

4. PC BAND–GAP DEVICES: WAVEGUIDES AND CAVITIES

4.1. Pc Waveguides

PC waveguides are one of the most important elements in PC integrated optics, because they offer efficient channels for light propagating at wavelength scale and connect different devices in integrated optical circuits. Usually, PC waveguides are formed by removing one row of holes in a PC structure. The line defects can generate defect states within the complete PBG and serve as waveguide channels for light to propagate efficiently and freely in PC structures. Among many 2D PC slab structures, the triangular lattice of air holes has a relatively large band gap for TE-like electromagnetic modes, where the magnetic field points in the perpendicular direction while the electric field is dominantly within the lateral plane of the slab. In addition, the structures allow for easy fabrication by standard planar nanofabrication technologies such as FIB and EBL, and have good mechanical stability. For these reasons, they are widely and dominantly adopted in designing and exploring PC based integrated optical devices [35,36].

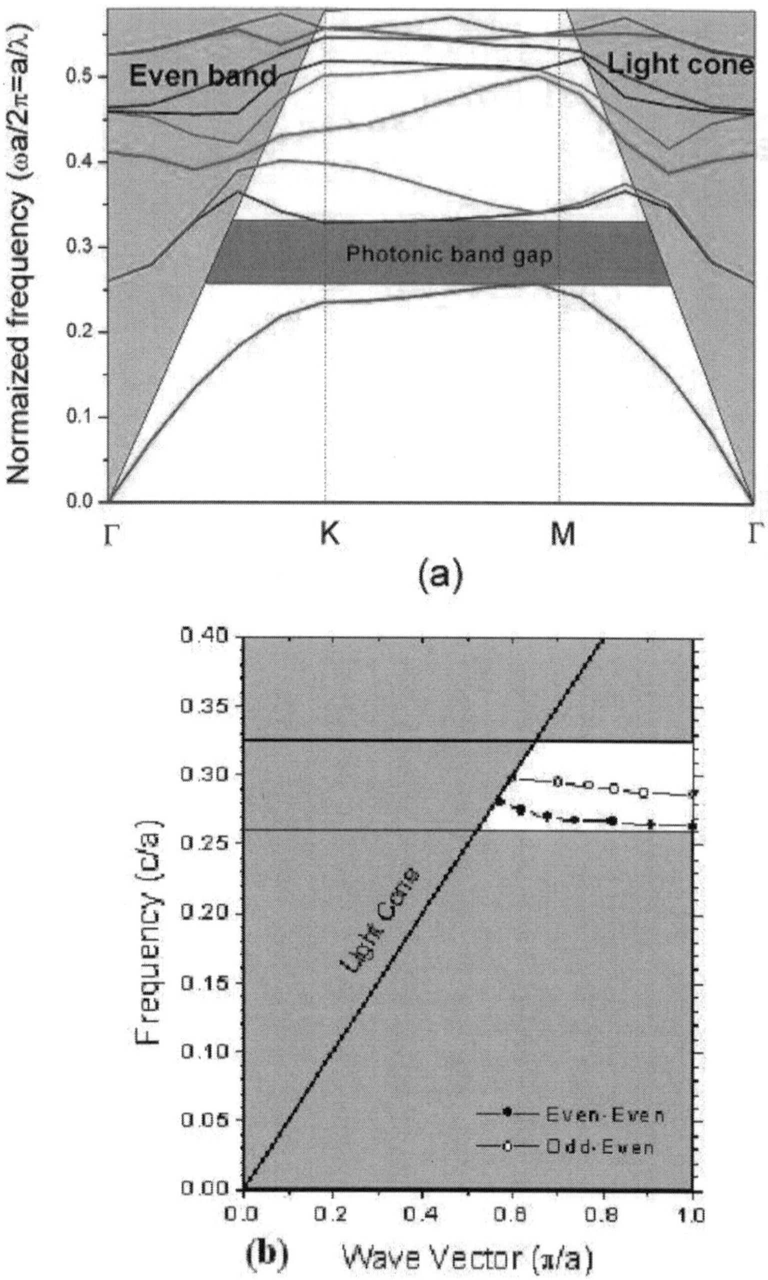

Figure 3. a) Photonic band structures for air holes triangular lattice PC slab; (b) band diagrams for a PC W1 waveguide, where one row of air holes is removed along the Γ−K direction. The upper and lower bands correspond to the even-symmetric and odd-symmetric guided mode, respectively.

Figure 3(a) shows the calculated TE-like mode photonic band structures of a particular 2D triangular lattice PC slab, where a wide complete PBG is clearly seen. In the region, light propagation inside the PC is prohibited. When removing one or several rows of air holes in the PC structures, some allowed modes (defect states) appear within the PBG [Fig. 3(b)], and they can be used to create waveguides or cavities. In most works, single-mode or multi-mode optical waveguides are usually made along the Γ−K direction in the triangular lattice PC. It has been well established in plenty of literatures that the number of waveguide modes as well as the width of the transmission windows can be controlled by tuning the core width of the line defects. However, waveguides along other directions in the triangular lattice PC were rarely discussed. Just like the Γ−K direction, waveguides along the Γ−M direction should also be able to guide confined modes due to the existence of a complete PBG in the 2D triangular lattice PC. By removing a line of diamond areas, we can obtain a cluster-like waveguide along the Γ−M direction as depicted in Fig. 4(a). This kind of waveguide is called Γ−M waveguide [37].

Figure 4(b) is the SEM picture of the original Γ−M waveguide, where the air holes remain to locate at the original lattice site and the radius of all air holes remains the same. The air holes are directly drilled by FIB and the lattice constant is 430 nm. According to our simulation and experiment results, we find that the width of the propagation modes for the original Γ−M waveguide [Fig. 4(b)] is only 22 nm. Then we optimize the geometry to improve its transmission characteristics. We shrink the radius r1 of the air holes in the two nearest-neighboring rows around the waveguide and enlarge the radius r2 of the air holes in the two second-nearest-neighboring rows, as shown in Fig. 4(c). The key point is to generate a transport pathway with walls as smooth as possible. According to our simulations, the parameters corresponding to an optimized waveguide are that r1 = 50 nm and r2 = 170 nm, while the radius of the original air holes is r0 = 120 nm. Figures 5(a) and (b) are the calculated dispersion relations of the original and optimized Γ−M waveguide, respectively. It's shown that the optimized waveguides have a high pass band that is much broader than the original waveguide. We can also obtain the same conclusion from the measured transmission spectra in Figs. 5(c) and (d). Besides, the intensities of the transmission spectra are much higher than the original one. As the Γ−M waveguide is perpendicular to the usual Γ−K waveguide, it offers an alternative to construct a waveguide interconnection beyond the usual scheme of Γ−K with Γ−K waveguides. A high-performance wide-band Γ−M waveguide should be of great help to build integrated-optical devices, such as interconnection networks, channel-drop filters, and wave division multiplexers, with more flexible geometrical configurations in 2D PC slabs.

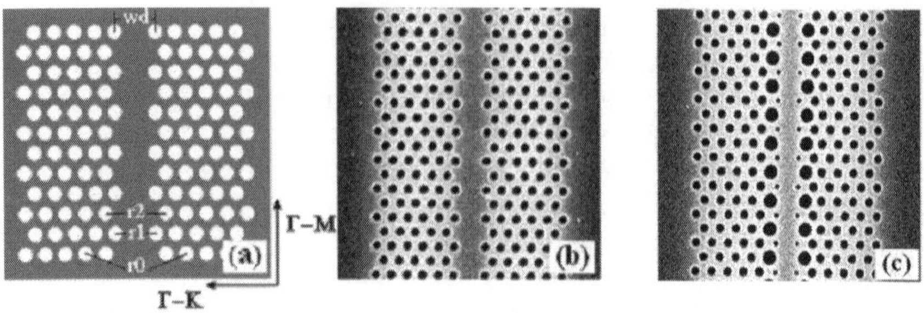

Figure 4. a) Schematic of Γ−M waveguide constructed in a triangular-lattice PC slab. The width of the waveguide wdwd, as well as the radius of air holes in the first and second row r1 and r2 , are the three crucial parameters to optimize the width of the transmission windows; (b) and (c) are SEM pictures of original and optimized Γ−M waveguides [37].

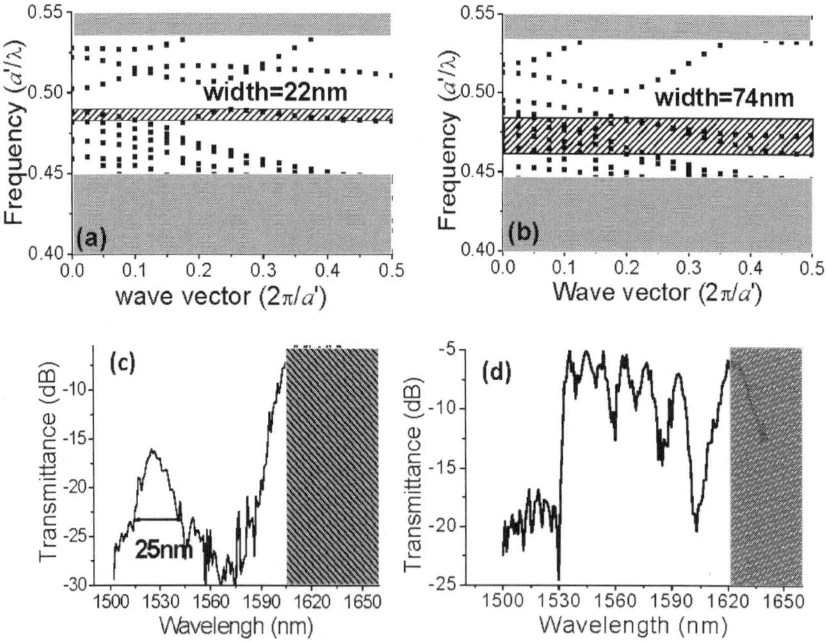

Figure 5. Calculated modal dispersion relation of (a) the original Γ−M waveguide and (b) an optimized Γ−M waveguide. The band width of the waveguide modes (within the dashed boxes) is 22 nm in the original waveguide, which has parameters: lattice constant a = 430 nm, hole radius r0=r1=r2=120 nm, and waveguide width w0=2a. After optimization by the following parameters as a = 430 nm, r_1 = 50 nm, r_2 = 170 nm, and w_d = $0.65w_0$, the waveguide band width is significantly broadened to 74 nm; (c) and (d) are the corresponding measured transmission spectra of the original and optimized waveguides [37].

Based on the design of an optimized Γ−M waveguide, we combine the Γ−K waveguide and Γ−M waveguide together to form a 90° waveguide bend as schematically depicted in Fig. 6(a) [38]. This is the first design of a 90° waveguide bend in the 2D triangular-lattice PC. The whole waveguide bend system is composed of two Γ−K waveguides as the input and output ports and the center Γ−M waveguide as the connection component. In other words, the structure involves two consecutive segments of 90° waveguide bends. In constructing the waveguide bend, we have used the optimized Γ−M waveguide discussed in the above and depicted in Fig. 4(c). The guided modes in the optimized Γ−M waveguide have better phase-matching and mode profile matching with the normal Γ−K waveguides. In the same time, we modify the bend corner geometry by fabricating smaller air holes in the corner to make the guided modes between the two kinds of waveguides matching better. Based on a serial of simulation and experiment tests, we find the best values for r1 and r2 . Figure 6(b) shows the SEM picture of the waveguide bend with r1 = 50 nm and r2 = 150 nm. After optimization, we get 70 nm pass band width while the transmission efficiency of a single bend is 45%. The proposed 90° waveguide bends can help to construct integrated optical circuits with more flexible and diversified infrastructures.

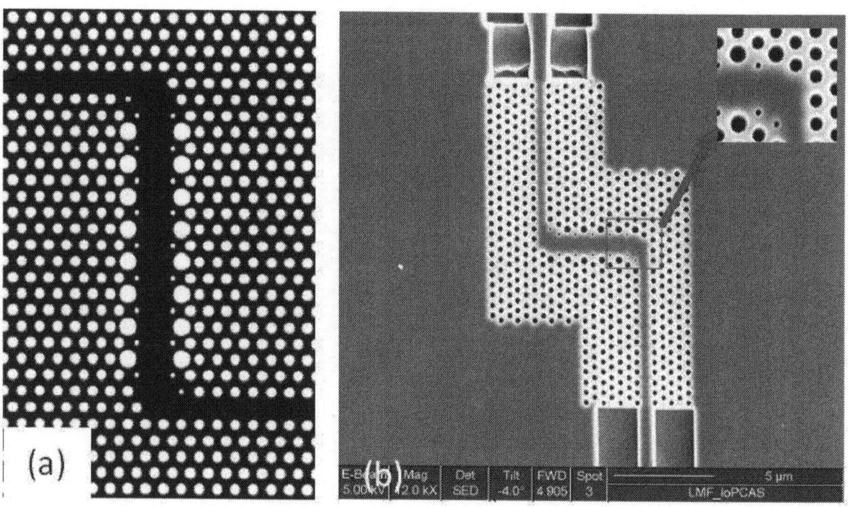

Figure 6. a) Schematic geometry of 90° waveguide bends in a triangular lattice PC slab with optimized Γ−M waveguide; (b) SEM picture of a practical sample of 90° waveguide bends with optimized bend corner geometry [38].

4.2. Coupled–Cavity Waveguide

Moreover, we have designed an air-bridged silicon PC coupled-cavity waveguides (PCCCWs) [39] and mapped its near-field optical distributions at different wavelengths around 1550 nm with the scanning near-field optical microscopy (SNOM) technology. For PCCCWs, the eigenmodes usually have relatively narrow bandwidth with slow group velocity in the whole band range. Previously, slow light propagation in such specific PCCCWs had still not been experimentally studied via SNOM technique.Figures 7(b)–7(f) show the calculated optical field distribution profiles at different wavelengths with a simulation model schematized in Fig. 7(a). We fabricated the PCCCW in the SOI wafers with FIB system. Figure 8(a) displays the SEM image of the element composed of the central PCCCW (encircled by a red square), two identical W1 PC waveguides, and the input/output ridge waveguides.Figure 8(b)–8(f) displays the near-field optical intensity distribution patterns of the PCCCW at different wavelengths. The scanning area is 12 μm × 15 μm with the incident light propagating upwards from the bottom of the image. Straight yellow lines in Figs. 8(b)–8(f) are used to label the position for showing the cross-sectional profiles of the field distribution patterns.

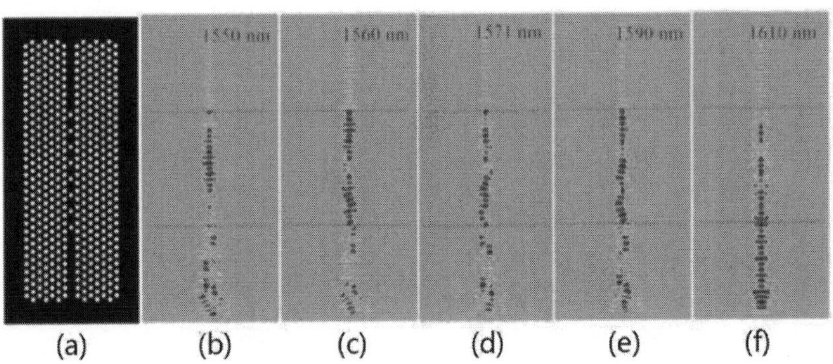

Figure 7. Simulation model (a) and calculated optical field distributions at (b) 1550 nm; (c) 1560 nm; (d) 1571 nm; (e) 1590 nm; and (f) 1610 nm [39].

The optical intensity distribution patterns are different at 1550 and 1610 nm, even though both of them mainly appear as a single narrow line along the central PCCCW region with a full width at half maximum (FWHM) of about 350 nm. Precisely speaking, the pattern demonstrates a little bit shoulder as a result of mode superposition at 1550 nm, since it comprises two eigenmodes. At 1610 nm, the calculated result consists well with the experimental one in Fig. 8(f), which presents a single line along the whole waveguide. The pattern appears bright and wide with obvious interference nodes in the PCCCW section. The simulated field distribution profiles in the W1 PC waveguide sections agree well with the detected ones at all these wavelengths, which show a snake-like/single-line profile in the input/output W1 PC waveguide except that of a snake-like profile in the output W1 PC waveguide at 1560 nm. In addition, the simulated field distribution patterns of the snake-like profile in the PCCCW section appear deviating greatly from the detected ones at 1550, 1571, and 1590 nm. However, if we calculate the optical field distribution patterns at 1550, 1560, and 1571 nm with the even-to-odd amplitude ratios of 1:4, 1:1, 1:4, and 1:6, respectively, we can find the simulated results are consistent with the experimental patterns evolving from single-line, to snake-like, and then to double-line structures for the PCCCW section. Combination of the near-field optical detection and theoretical simulation shows that SNOM is an efficient tool to study the optical propagation in the PCCCW and can help to design slow light elements.

4.3. High–*Q* Cavity

Quantum information processing and quantum state manipulation have received great attentions because of their potential revolutionary impact on future network communication. Optical cavities, which can be used to store information, are considered to be one of the most important devices in the quantum communication application. The generation and teleportation of qubits require sufficiently high value of Q/V, where V stands for the mode volume of the optical cavities. As a result, high-Qoptical cavities show great potential application in quantum information. Among all the optical cavities, 2D PC slab cavities are the best choice because of their simultaneous high-Q and small mode volume characteristics. It has been reported that Rabi splitting can be observed experimentally when the quantum dots are introduced into the PC cavities with high Q/V [40,41]. Moreover, due to the development of nanofabrication technique, multiple high-Q PC slab cavities can be fabricated at the same time on a single slab by the EBL and ICP etching technique. Once atoms or quantum dots are embedded into the high-Q PC cavities, various quantum phenomena can be demonstrated on chip. Recently, our works on high-Q silicon PC microcavities have achieved great progress after extensive exploration and delicate improvement of nanofabrication techniques and sample processing techniques have been made [42].

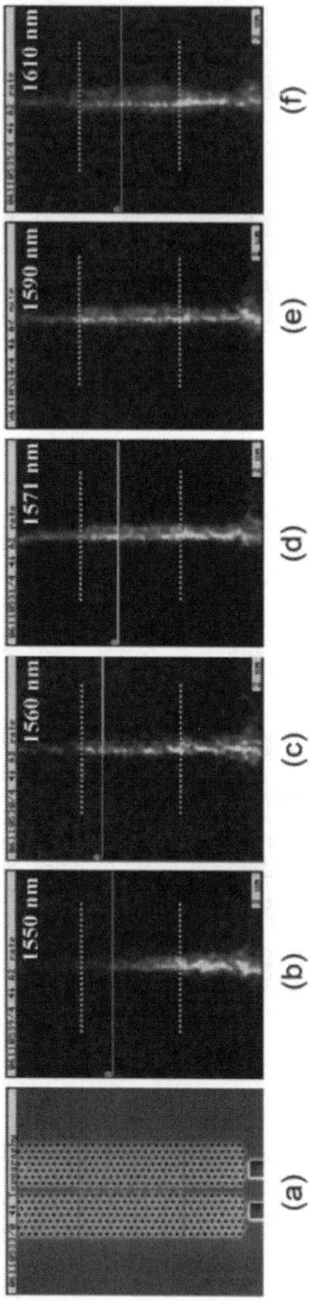

Figure 8. a) SEM topographic image, and the near-field optical intensity distributions at (b) 1550 nm; (c) 1560 nm; (d) 1571 nm; (e) 1590 nm; and (f) 1610 nm. The white dotted lines in each optical picture denote the interface between the W1 PC waveguide and PCCCW. All pictures were obtained for the same scanning area of 12 μm × 15 μm [39].

We focus on studying the L3 PC microcavities formed by removing three cylindrical air holes in the Γ−K direction in a triangular lattice [Fig. 9(a)]. The lattice constant is 430 nm, the radius of cylindrical air hole is 120 nm and the thickness of silicon slab is 235 nm. The FDTD calculation results indicate that the L3 PC microcavity possesses a quality factor of about 5300 [Fig. 9(c)]. After trying hundreds of simulations, we find that the positions of air holes at the edges of the microcavities affect the Q factor dramatically. The electric field pattern of the cavity mode can be tuned to be Gaussian-type by displacing the six air holes outwardly at the edges of the microcavities, and this can increase the quality factor significantly [41,42]. The optimal displacement is found to be 73, 10 and 73 nm for the first, second and third air holes at both edges of the microcavities, which is depicted in Fig. 9(b). The maximum quality factor of 127,323 [Fig. 9(d)] can be achieved, which is 20 times larger compared with the unadjusted one.

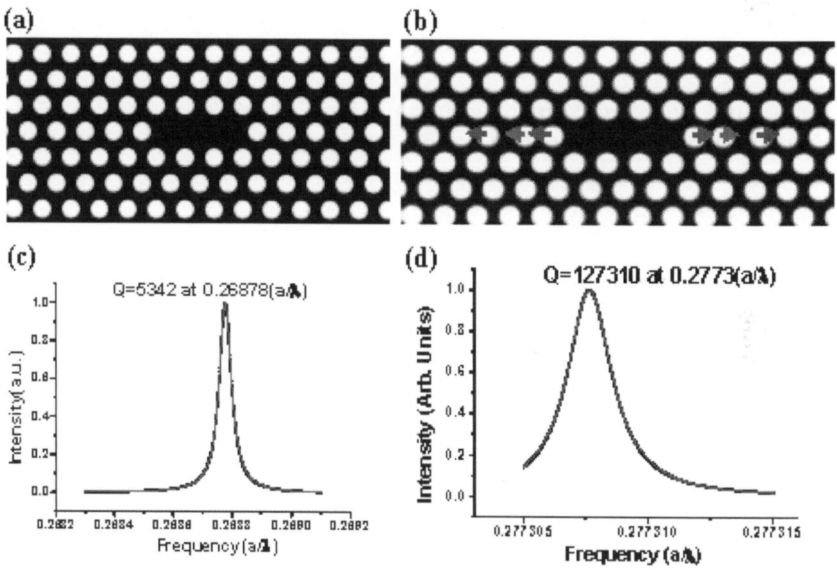

Figure 9. Schematics of (a) the original PC L3 nanocavity and (b) the optimized nanocavity; (c), (d) show radiation spectra of the original PC L3 nanocavity and the optimized nanocavity, respectively [42].

Based on the optimal parameters, we successfully fabricate the designed high-Q planar L3 PC microcavities in SOI wafer by implementing EBL and ICP [42]. As can be seen in Fig. 10(a), the L3 microcavity is side-coupled to a W1 waveguide with the barrier of three rows of air holes. The samples are measured by our home-made fiber coupling system as described in the above section. When the incident wavelength is off-resonant, light cannot couple with the microcavity, leading to strong output. While, at resonance most energy is tunneled into the microcavity, resulting in weak output. For the case of high-Q microcavity, a sharp transmission dip is expected in the transmission spectrum. The lattice constant, radius of cylindrical air hole and the thickness of silicon slab are 430, 120 and 235 nm, respectively. Limited by the fabrication accuracy of 10 nm, the displacement is adjusted to be 80, 20 and 80 nm for the six air holes at both edges of the microcavities. Figure 10(a) shows the enlarged view of the cavity region. A sharp and narrow transmission dip is observed at the 1567.35 nm in the measured transmission spectrum [Fig. 10(c)]. For the purpose of extracting the quality factor accurately, we finely tune the wavelength between 1565 and 1570 nm. The measured spectrum is illustrated in Fig. 10(d) and the Q-factor as large as 71,243 is obtained. Nevertheless, there are some deviation between the simulation and experiment. For example, the resonant wavelength is 16.75 nm red-shifted from the simulated result and the maximum quality factor is significantly less than the calculated value of 127,323 [Fig. 10(b)]. We believe that the deviation is caused by the imperfection of the cylindrical air holes and the actual radius is not exactly the same as the value in simulation. The success of fabricating high-Q silicon PC slab microcavities enables us to investigate various interesting quantum phenomena, such as strong coupling between light and quantum system, quantum information processing technique, single photon source, all-solid quantum manipulation and high-quality biochemistry sensing devices.

4.4. Channel Drop Filters

Channel drop filters are key components for extraction of light trapped in a point-defect cavity to a neighboring waveguide and they sit on the basis of wave-division multiplexers and demultiplexers. They have great applications in a wide variety of fields, such as photonic integrated circuits, telecommunications, and quantum informatics. Based on the simulation and experiment experiences about PC waveguides, we design and fabricate an ultra compact three ports filter in 2D air-bridged silicon PC slab by closing the bus waveguide for 100% reflection feedback. Figure 11(a) shows the SEM picture of the three ports filter structure [43]. This filter was fabricated by EBL and ICP techniques. The lattice constant of the PC and the radius of the air hole are 430 and 145 nm, respectively. Port 3 is the input waveguide channel, while ports 1 and 2 are two output waveguide channels, respectively. They are formed by missing one row of air holes along the Γ−K direction of the triangular-lattice PC, the so called W1 waveguide. C1 and C2 are two point-defect cavities. The distance between the center of the defect cavity and the neighboring waveguide is 3 rows of holes in the y direction. As seen in Fig. 11(a), the C1 cavity consists of three missing air holes, and the two air holes at the cavity edges are shifted outward by 10 nm apart from the regular positions. Similarly, those of the C2 cavity are shifted by 20 nm. The slight shift of air holes is conducive to confine light inside the cavity and leads to a higher quality factor. Meanwhile, the different shifts of the two cavities make the resonant wavelengths slightly different. The experiment results [Figs. 11(b) and (c)] show that the resonant wavelengths of C1 and C2 are 1529.5 and 1531 nm, respectively. The wavelength spacing of the two cavities is about 1.5 nm and might be further reduced by continuously changing the size of the cavity. The full widths at half maximum of the peaks are 1.5 and 1.4 nm and the corresponding quality factors are about 1020 and 1090, respectively. To estimate the drop efficiency, a reference straight waveguide of the same parameters is positioned near the three-port filter. By keeping the same intensity of input light, the transmission intensities of the reference waveguide and port 1 are 0.330 and 0.158 μw, respectively, when the input wavelength is set at 1529.5 nm. The drop efficiency of port 1 is roughly estimated to be 48% and a similar result has been obtained at port 2.

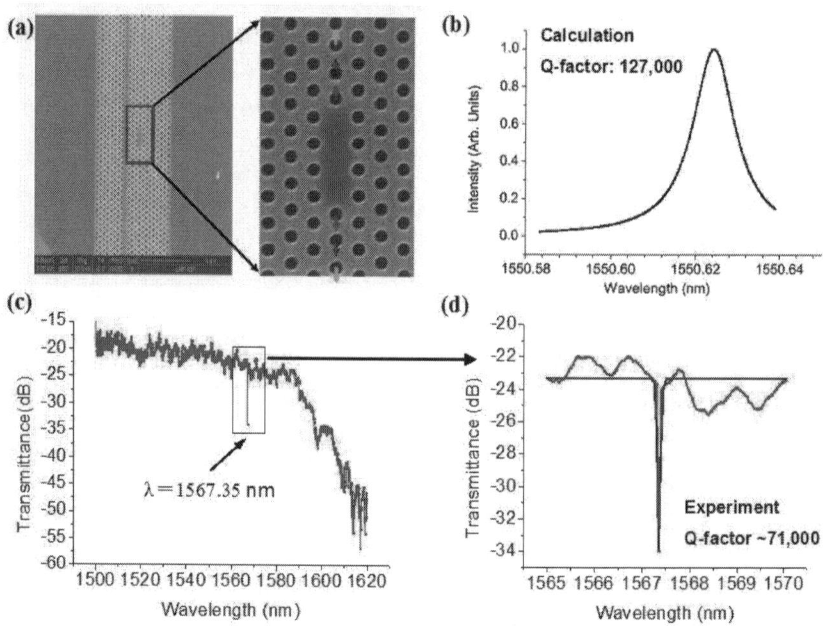

Figure 10. a) SEM pictures of one of the fabricated samples, including the L3 nanocavtiy with displaced air holes; (b) radiation spectra calculated by FDTD method; (c) and (d) show transmission spectra of one of the fabricated samples. The maximum Q value of up to 71000 is obtained [42].

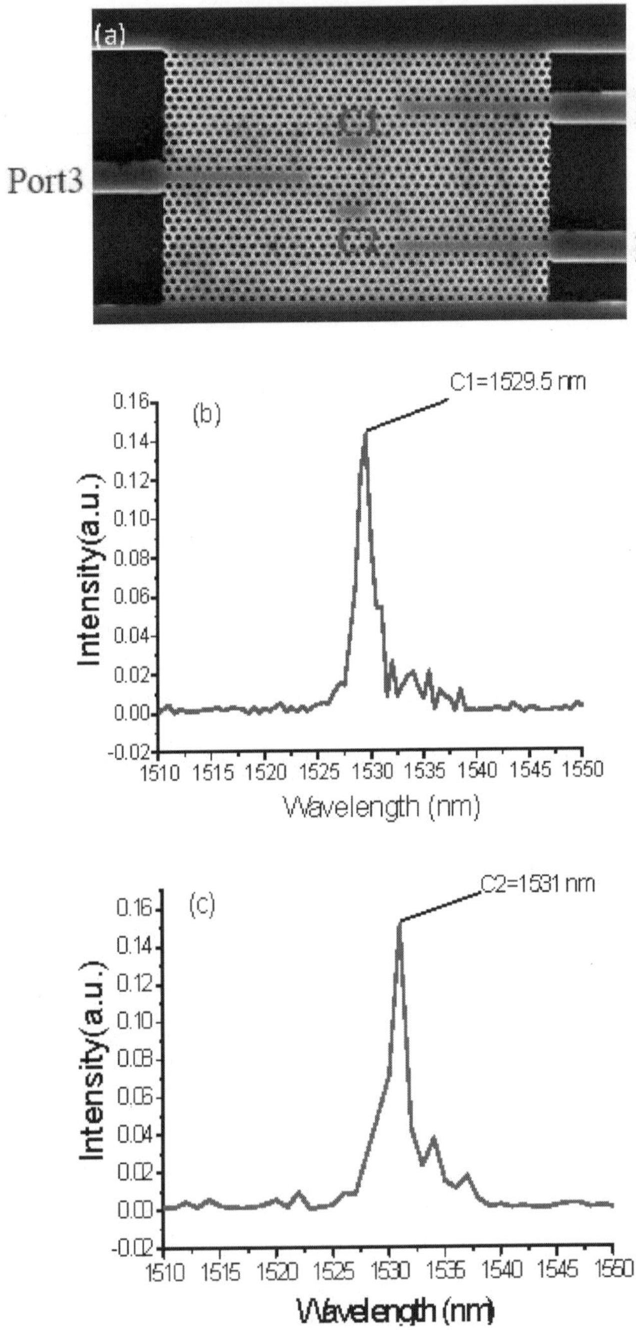

Figure 11. a) SEM image of the three-port filter; (b) and (c) the measured transmission spectra at ports 1 and 2, respectively [43].

It has been well known that structure is the kernel of filter design. Usually, the regulation of microcavity resonant frequency is obtained by changing the size of the cavities. We have proposed a new way to design multi-channel filters by changing the shape of the air holes [44]. When the shape of the air holes changes from circle to ellipse, two parameters, the ellipticity and the orientation angle of the ellipse, in addition to its size can be further explored and they can have a great influence on localized cavity modes. Therefore, we can use this for some special purpose.

Figure 12(a) schematically shows a one-channel PC filter. A horizontal channel (W1 waveguide) serves as the input signal channel, which is created by removing a single line of air holes along the Γ−K direction. A cavity is formed by removing three air holes along the Γ−K direction that is rotated 60° from the W1 waveguide. It is located four rows away from the major channel and is connected with the major channel through an indirect side coupling. Another single-mode waveguide is formed parallel to the cavity and serves as the output signal channel. Figure 12(b) shows an enlarged picture of the filter in the region around the cavity. One of its axes is oriented counterclockwise by an angle θθ to the x axis, namely, the input light propagation direction. The sizes of the axes parallel and perpendicular to this orientation are a and b, respectively. Now we have great structural freedom to tune the optical properties of the new PC filter by changing the parameters of θ, a and b. To show this point, we design and fabricate a four-channel PC filter by using different cavity parameters as described in Table 1. The SEM picture of the fabricated four-channel filter is displayed in Fig. 13(a). Four cavities are located on the two sides of the central linear W1 waveguide. They are engineered by leaving several air holes unetched in the Γ−K orientation. The input signal propagates upwards from the bottom input ridge waveguide. Each cavity is coupled with another W1 waveguide that is connected to a ridge waveguide, which serves as the output signal channel.

The simulation and experimental results of the transmission spectra for the four channels are displayed in Figs. 13(b) and (c). Although significant noise exists, a resonant peak can be clearly found for each channel. The peaks are located at 1549, 1541, 1567, and 1560 nm for channels 1, 2, 3, and 4, respectively. The results confirm that the air-hole shape has a great influence on the functionality of the PC filter devices. The elliptical air holes can induce a fine tuning of the resonant wavelength by changing the ellipticity of the elliptical air holes.

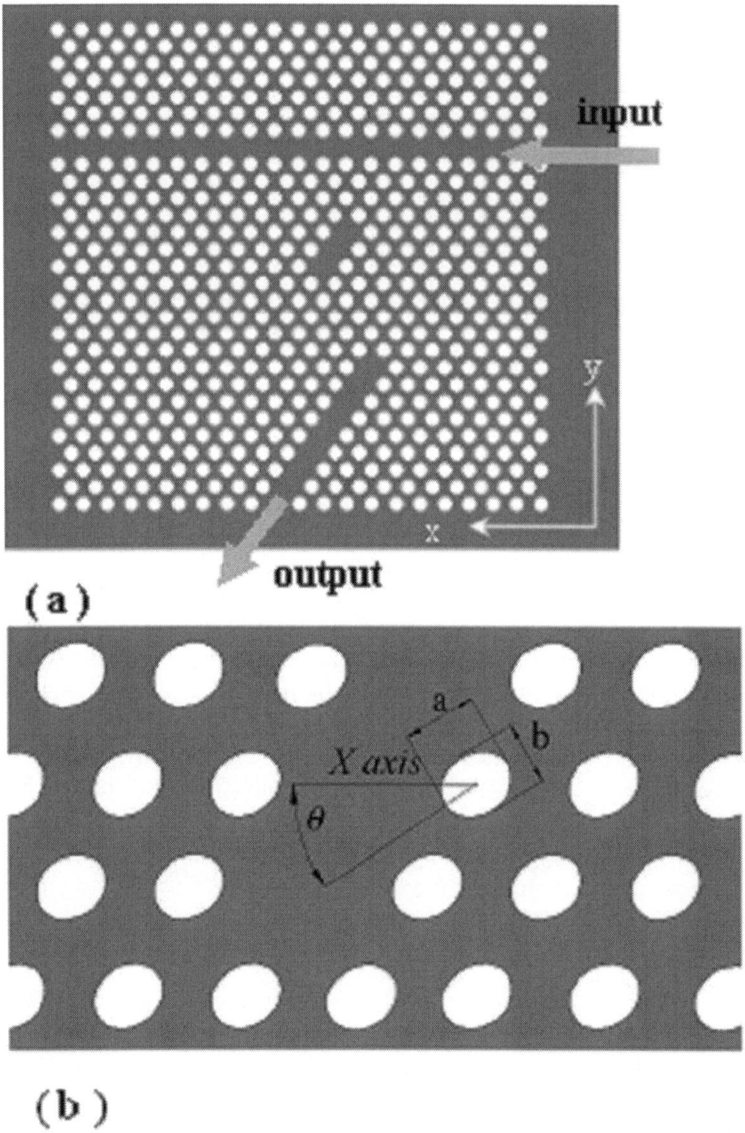

Figure 12. a) Schematic view of a one-channel PC filter, the major channel lies in the *x* direction, and the cavity and output side channel are parallel to the Γ−K direction of the triangular lattice; (b) enlarged view of the filter around the cavity. The air holes have a general elliptical shape with one of its axes oriented counterclockwise by an angle θ with respect to the x axis. The two axes are of size a and b, respectively [44].

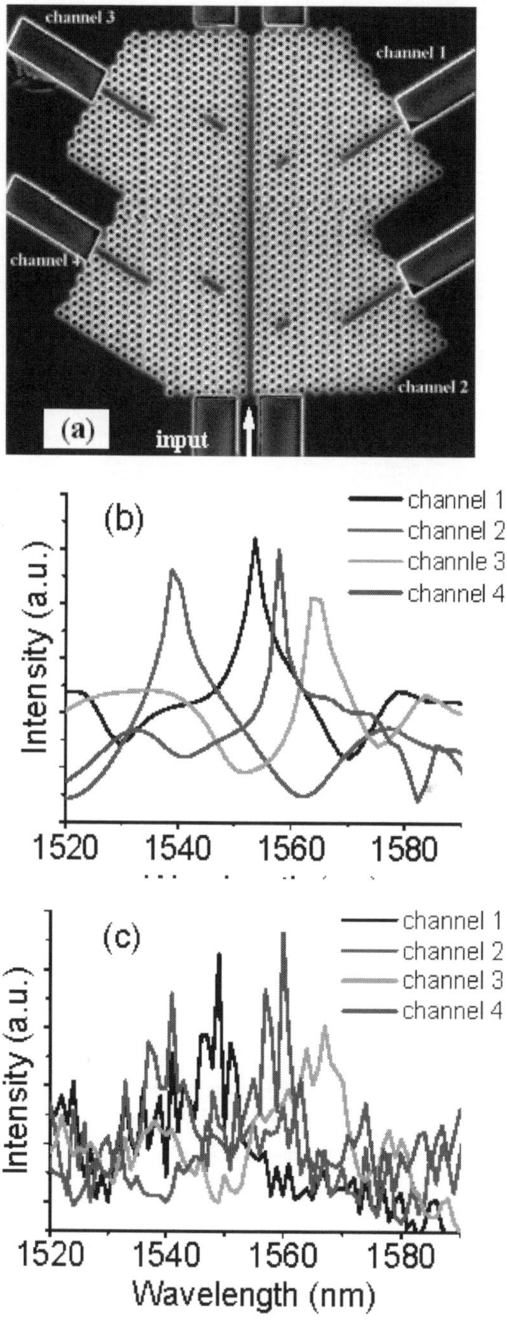

Figure 13. a) SEM image of the four-channel filter; (b) Simulation results of transmission spectra for the four-channel filter. (c) Experiment results of transmission spectra for the same filter [44].

Table 1. Structural parameters in the four-channel filter

channel	lattice constant /nm	number of missing air holes in cavities	long axis *a* /nm	shot axis *b* /nm	angle θ	theoretical resonant peak /nm	measured resonant peak /nm	deviation nm
1	420	2	240	200	0	1553	1549	4
2	430	2	260	240	0	1539	1541	2
3	420	3	240	220	0	1563	1567	4
4	430	3	280	240	0	1558	1560	2

As described in the preceding section, a high-performance wide-band Γ−M waveguide can be formed by modifying the radii of the air holes along the pathway. The combination of Γ−M waveguides and Γ−K waveguides can offer a more flexible way to interconnect and couple between different devices. We have proposed a type of PC filter using these two kinds of waveguides [45]. The Γ−M waveguide and Γ−K waveguide are used as the input and output signal channels respectively, and they are connected via cavity resonance. Figure 14(a) shows the SEM picture of a four-channel filter structure. We change the size of the cavities by moving the end points of the cavity [marked with "a, b" and "c, d" as shown in the inset of Fig. 14(b)] to change the resonant frequency. Table 2 gives the detailed parameters of the four cavities. The experiment demonstrates that the four resonant peaks are at the wavelengths of 1543, 1545, 1548 and 1551 nm, as shown in Fig. 14(b). In spite of the slight shift in the resonant peak toward higher frequency, which we believe is induced by the uncertainties in the fabrication, the experimental results are in fairly good agreement with the simulation results, where the maximum relative deviation of resonant wavelength is within 2 nm. These results clearly demonstrate that the designed Γ−M waveguides can act together with the usual Γ−K waveguide to construct high-performance multichannel filers with more structural flexibility. In our experiment, we also use the CCD camera to directly monitor the transport of infrared signal within the channel-drop filter. The situation of on-resonance and off-resonance can be clearly visualized and distinguished from the CCD camera images. One typical case is shown in Fig. 14(c).

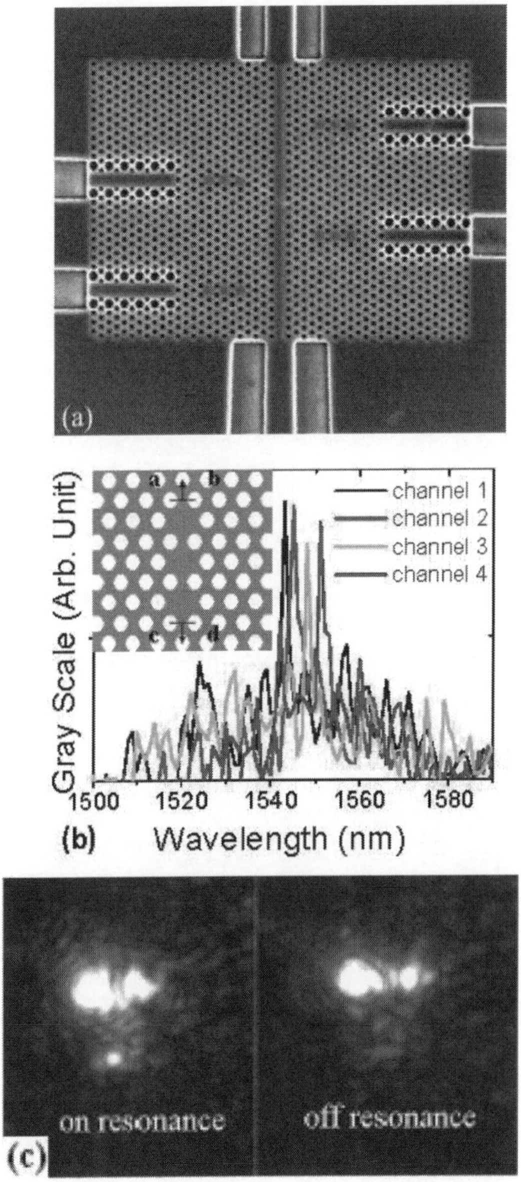

Figure 14. a) SEM image of the fabricated four-channel filter. Four cavities are located on the two sides of the input waveguide; (b) experimental transmission spectra of the four channel filter in linear scale. The inset picture illustrates two groups of end points (air-hole centers) of the cavity marked with "a, b" and "c, d." Black arrows indicate the moving direction of these air holes; (c) infrared CCD camera imaging of the output signal observed in experiment for one channel of the sample. A bright spot appears at the end of the output channel when the input wavelength coincides with the resonant wavelength and disappears when it is at off-resonance [45].

Table 2. Structural parameters in the four channel Γ−M and Γ−M waveguides filter

channel	end points moving distance /nm	theoretical resonant peak /nm	measured resonant peak /nm	theoretical distance from channel 1 /nm	experimental distance from channel 1 /nm	deviation /nm
1	0	1550	1543	-	-	-
2	5	1551.5	1545	1.5	2	0.5
3	10	1553	1548	3	5	2
4	15	1556	1551	6	8	2

5. PC BAND–ENGINEERING DEVICES FOR ANOMALOUS TRANSPORT CONTROL

In previous sections, we discuss several PC devices, including waveguides, cavities, and channel-drop filters that are built on the silicon 2D PC platform. These devices work on defects that are brought into PBG and they can be considered as PBG materials. As we have mentioned, PC structures possess another important feature: photonic pass bands. In this section we show several example devices that implement the dispersion and refraction properties of PCs at their transmission bands.

Let's first make an overview of how band dispersion engineering works. Figure 15(a) shows the typical photonic bands structure of an air-bridge PC slab structure composed of a square-lattice array of air holes etched in silicon slab. The areas labeled in Fig. 15(a) show the unconventional light propagation, self-collimation and negative refraction, in PC. One effective way to understand and exploit desirable light propagation properties in PC is using the equifrequency surface (EFS) contours, as shown in Figs. 15(b) and 15(c). Figure 15(b) shows the EFS contours of the first TE-like band. The EFS contours in the red line frame are flat, meaning that this is the self-collimation region. The reason is that the group velocity, which is parallel to the gradient of the EFS, is pointing in the same direction for all the modes located within the region. As a result, if light propagates along the Γ−M direction, it does not suffer any diffraction in the PC. $\Gamma=(0,0)(\pi/a)\Gamma=(0,0)(\pi/a)$ and $M=(1,1)(\pi/a)M=(1,1)(\pi/a)$ are high-symmetric points in the first Brillouin zone for square lattice. This kind of PC structures can be used as the channelless waveguide in integrated optic devices. Figure 15(c) shows EFS contours of the second band. The EFS contours are roughly circular around the direction at the reduced frequency ($a/\lambda a/\lambda$) range 0.28−0.31, as indicated in the red solid line frame. When the frequency increases, the EFS contours move toward the Γ point, which indicates the existence of negative refraction in the region. If the incident light is mainly parallel to the Γ−M direction, the PC will behave like an isotropic medium with a negative index of refraction in that particular frequency range.

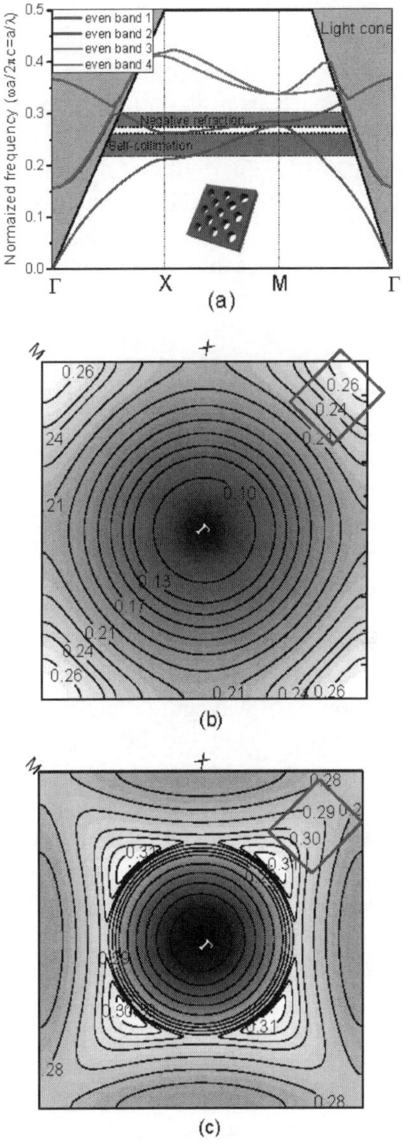

Figure 15. a) Photonic band structures of TE-like bands for an air-holes square-lattice PC slab; (b) EFS contours of the TE-like first band for the same PC show that self-collimation can occur in the direction around the Γ−M direction; (c) EFS contours of the TE-like second band show that negative refraction can occur in the direction around the Γ−M direction [46].

5.1. Negative Refraction

Based on the above analysis, we designed and fabricated an air-bridged PC structure that exhibited negative refraction of infrared light [46]. The structure is schematically shown in Fig. 16(a). These structures are directly drilled by FIB technique. The input infrared signal channel is a silicon wire waveguide, which is inclined with respect to the surface normal by 10°. The lattice constant a of the square array is 460 nm and the diameter of the air hole is 220 nm. We first use 3D FDTD method to simulate the electromagnetic field intensity distribution at wavelength 1503 nm. The result is displayed in Fig. 16(b). We find strong reflection and scattering at the interface between the input waveguide and the PC structure. This is induced by the serious impedance mismatch at the interface, although the high index contrast air-bridged structure can achieve good optical confinement. To surpass this obstacle, we use a tapered air-holes connection layer at the input surface of PC structure to reduce the reflection and scattering losses. As shown in Fig. 16(b), a large fraction of light power from the input waveguide is coupled into the PC structure and negative refraction of light beam within the PC structure is clearly seen. Besides, the reflection or scattering of light at the input interface of the PC is very much reduced. This clearly indicates that the designed tapered interface can reduce the interface impedance mismatch remarkably. The calculated value of negative refraction angle is −45°.

In our measurement, TE-polarized light from a tunable semiconductor laser (1500−1640 nm) was first launched into a tapered single mode fiber, and then coupled to the silicon wire waveguide, and finally incident on the PC structure. The ordinary way to see the light propagation behavior is to directly observe the pattern of the radiated light from the top of the sample using a conventional microscopy objective and an infrared CCD camera. The result is shown in Fig. 16(c). The light spot at the middle bottom part of the pattern is the radiated light from the input silicon wire waveguide. The big light spot at the center represents the scattered light at the interface between the input wire waveguide and the PC due to impedance mismatch. There is also a small bright spot at the top right corner of the pattern, and it is recognized to result from the radiated light when the negative refraction beam hits the end facet of the PC structure. Because the TE-like modes are strongly confined guided mode on the silicon slab and the surface fields are nonradiative and evanescent with respect to the vertical direction of the PC slab, the far-field pattern observed and recorded by the ordinary optical microscopy is not able to reveal the detailed process about how the negative refraction beam propagates inside the PC structure unless the scattering of light by roughness and irregularity is sufficiently strong on the beam propagation path. This is indeed the case for Fig. 16(c). In fact, the small bright spot could not appear without the intentional introduction of the air slot at the far end of the PC structure. It would not be possible to tell which way the infrared beam would refract if without the aid of this scattering light spot.

In order to observe clearly and tell unambiguously the ray trace of the negative refraction beam in the PC structure, we used the SNOM technology (SNOM-100 Nanonics, Israel). A probe scans in the vicinity of the surface of the PC structure and records the near-field intensity distribution. The tip has a resolution of about 100 nm, i.e., 1/15 of the wavelength. The signal is recorded by an infrared single-photon detector, which allows us to capture very weak infrared signals. The probed near field information directly reflects light propagation properties of the TE-like modes for the PC and enables one to visualize the ray trace of the negative refraction light beam because the near field at the surface is an integral part of the modal profile of the confined guided modes that exponentially decay away from the surface of the slab. In the SNOM picture [Fig. 16(d)], a bright spot also appears at the front interface of the PC structure, but it is much smaller than the one in Fig. 16(c). The ray trace of the incident light beam along the silicon wire waveguide and its propagation along the negative refraction direction inside the PC structure can be clearly seen. The negative refraction angle is about −45°, which is in good agreement with the FDTD simulation presented in Fig. 16(b). The SNOM detection unambiguously discloses the negative refraction property of the designed PC.

On the other hand, ordinary positive refraction only occurs for TM-like confined modes, so the designed PC structure can behave as an efficient beam splitter in an integrated optical circuit. The high-resolution SNOM technology can greatly help one to directly visualize the ray trace and acquire deeper understanding on various anomalous wave propagation behaviors, such as super-prism, superlensing, self-collimation, and slow light in deliberately designed 2D PC slab structures in the optical wavelengths. This in turn can help to explore a wider regime of controlling light behaviors on the nanoscale for future basic science and high technology applications.

5.2. Self–Collimation Effect

Self-collimation effect is the propagation of light without diffraction along the propagation direction. This phenomenon has been used to construct non-channel waveguides, beam splitters and beam combiners [47,48]. The behaviors of these devices are determined by the performance of the self-collimation effect. Recently we have designed and realized a simple structure composed by a square lattice array of elliptical air-holes where broadband large-angle self-collimation effect is observed for TE-like guided modes in infrared wavelength [49].

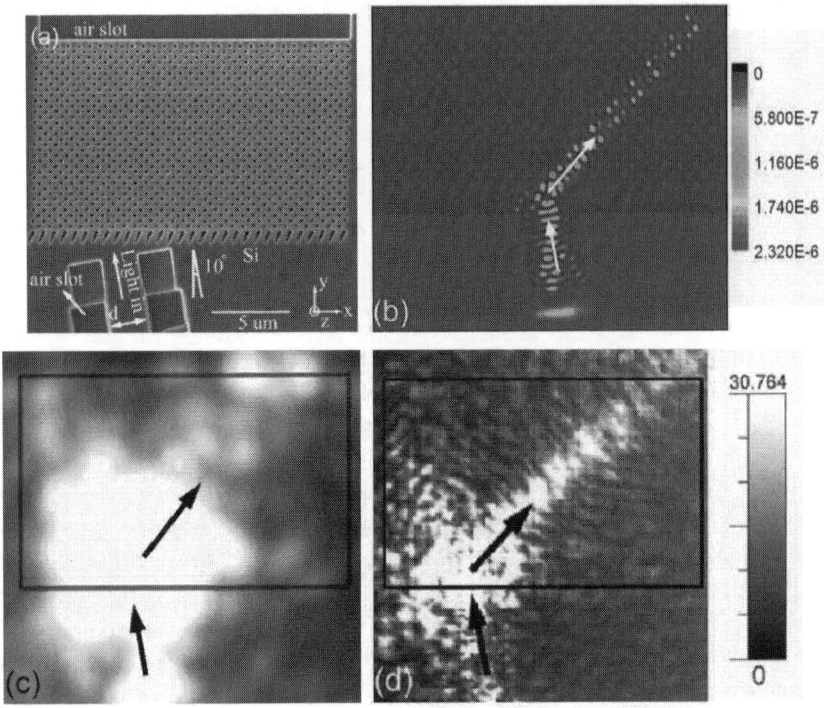

Figure 16. a) SEM picture of the PC structure and an input waveguide. The width of the waveguide d is 2 μm; (b) Light intensity distribution of TE-like modes for PC with deliberately designed tapered air-holes interface; (c) Directly observed pattern of the radiated light of λ=1500 nm from the top using an objective lens; (d) SNOM picture of the negative refraction of the same wavelength. In each picture, the boundary of the PC structure is superimposed as solid lines [46].

Figure 17(a) shows our PC structure formed by a square lattice of elliptical holes. The calculated TE mode photonic band diagram of the fourth, fifth and sixth bands are shown in Fig. 17(b). The self-collimation effect can be observed at the gray regions within a broad normalized frequency range 0.36−0.390.36−0.39 and 0.43−0.460.43−0.46. For simplicity, we only consider the EFS contours of the fifth TE band [Fig. 17(c)]. The contours are flat at the normalized frequency between 0.43 and 0.46 for all the values of ky and not just in the vicinity of ky=0. This feature indicates that our structure can support self-collimation for incident light beams with large incident angles. Then the FDTD simulation by using the MEEP package is performed to verify our prediction. A structure with the size of 30 aa× 45 aa is considered. A Gaussian beam with a width of 4*a* propagates into the surface at 0°, 20° and 60° incident angles (Fig. 18). For simplicity, we only consider the minimum (0.36) and the maximum normalized frequency

(0.43). Figure 18 shows the electric field intensity distribution in the xy plane with 0°, 20° and 60° incident angles at normalized frequency .36 [Fig. 18(a)] and 0.46 [Fig. 18(b)]. From the simulation results, we find that the light beam is collimated along the propagation direction for each situation. However, the couple efficiency of the incident light becomes lower and lower with the increase of the incident angle. We do not show the field distributions for those incident angles that are larger than 60°. These six situations in Fig. 18 are sufficient to show the broadband large-angle characteristic of the self-collimation effect.

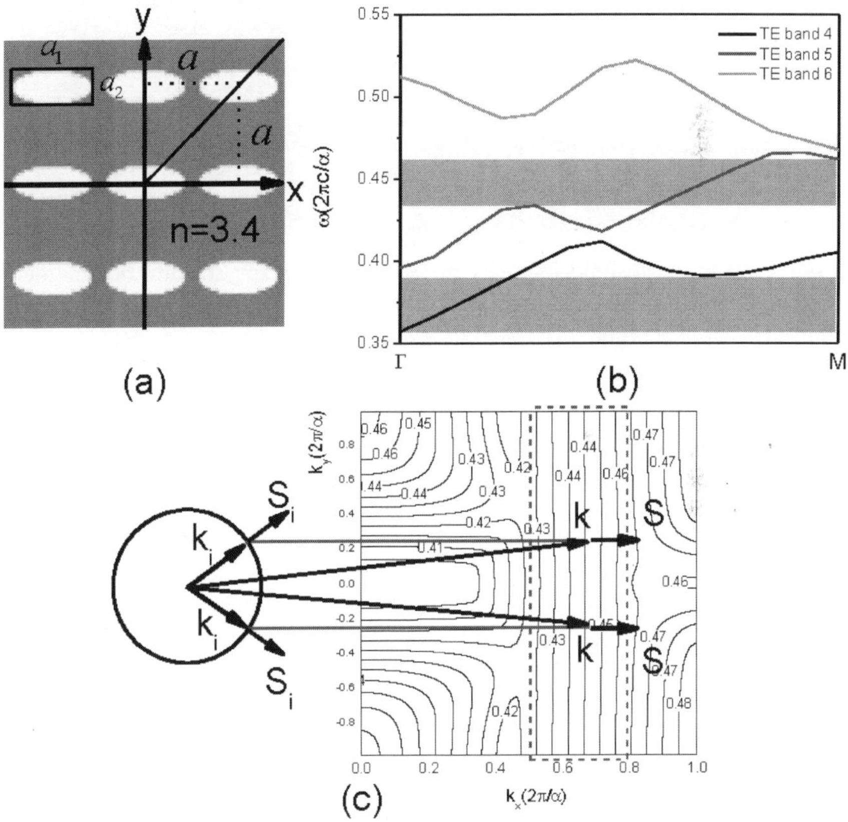

Figure 17. a) Schematic of the PC structure formed by a square lattice of elliptical holes; (b) band diagram of the fourth, fifth and sixth TE bands; (c) EFS contours of the fifth TE band [49].

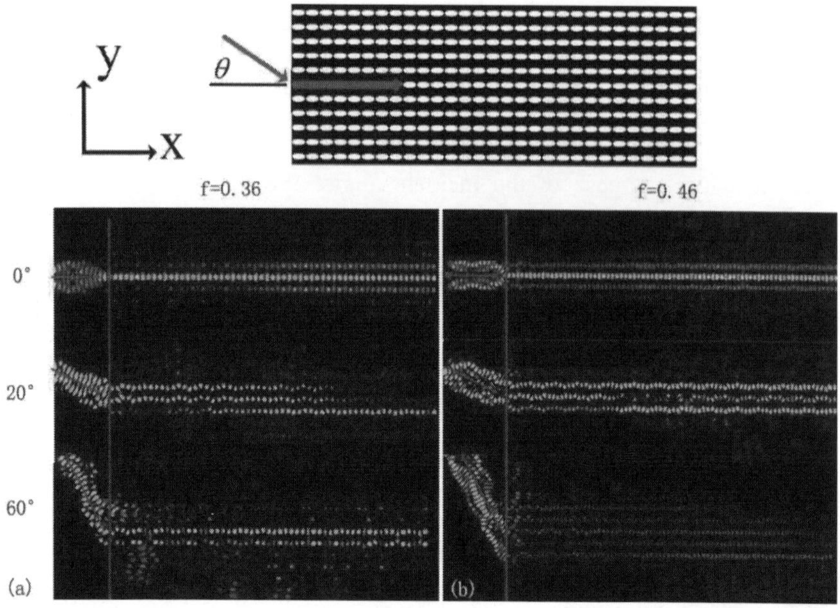

Figure 18. Electric field intensity distribution with 0°, 20° and 60° incident angles at the minimum normalized frequency 0.36 (a) and the maximum 0.46 (b). A FDTD method is used in the simulations [49].

Following our simulation results, we fabricate our PC structures in SOI substrate by EBL and ICP etching process. Figure 19 shows the SEM pictures of the designed PC structures with 0°, 20° or 60° incident waveguides. Ray trace of light beam is observed using IR camera and a high numerical aperture (NA = 0.50) objective. Detailed images of the field intensity of the scattered light are recorded for 0°, 20° and 60° incident angles for different incident wavelengths. Here we only show the patterns of the minimum and maximum wavelengths for each incident angle. They demonstrate strong light confinement along the propagation direction for all the situations. The experimental results are in good agreement with FDTD simulations. We believe that this kind of structure may have potential applications in beam combiners and multiplexers.

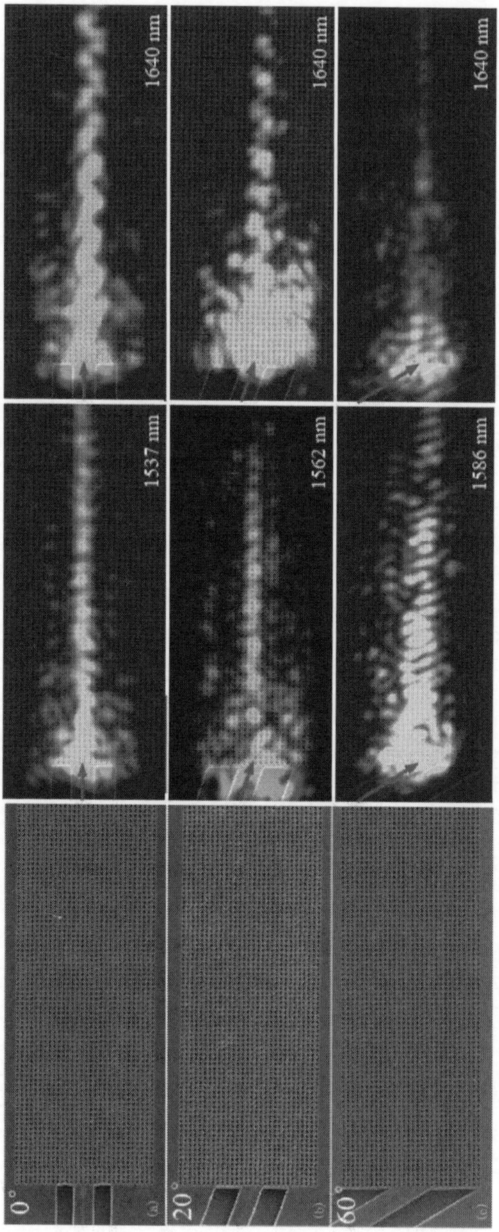

Figure 19. Left panels: SEM pictures of designed PC structures with 0° (a), 20° (b) and 60° (c) incident waveguide. Middle and right panels: Ray trace of light beam observed using IR camera and a high numerical aperture (NA = 0.50) objective. The patterns of the minimum and maximum wavelengths are shown for each incident angle [49].

6. ON–CHIP WAVELENGTH–SCALE OPTICAL DIODE AND ISOLATOR

Optical isolation is a long pursued object with fundamental difficulty in integrated photonics. The need to overcome this difficulty is becoming increasingly urgent with the emergence of silicon nano-photonics, which promises to create on-chip large-scale integrated optical systems. Motivated by the one-way effect, considerable effort has been dedicated to the study of unidirectional nonreciprocal transmission of electromagnetic waves, showing important promise in optical communications. Until now, on-chip integration of optical diode still stays in theory, particularly in silicon. These "optical diodes" include fluorescent dyes with a concentration gradient, absorbing multilayer systems, and second harmonic generators with a spatially varying wave vector mismatch. An electro-tunable optical isolator based on liquid-crystal heterojunctions, showing nonreciprocal transmission of circularly polarized light in photonic bandgap regions, has been reported. In another configuration using liquid crystals, linearly polarized light is used. In addition to many attempts on magneto-optical materials, optical isolators have also been fabricated using nonlinear optical processes and electro-absorption modulators.

An efficient routine to create optical diode is via time-reversal symmetry breaking or spatial inversion symmetry breaking [50], which could lead to the optical isolation in any device where the forward and backward transmissivity of light is very much different. We have reported a method for making unidirectional on-chip optical diodes based on the directional bandgap difference of two 2D square-lattice photonic crystals comprising a heterojunction structure and the break of the spatial inversion symmetry. Simulations confirm the existence of a clear isolation effect in the designed heterojunction structure. We fabricate these on-chip optical diodes in silicon and the near-infrared experiment results show high-performance optical isolation, in good agreement with the theoretical prediction [51]. This device may play the same basic role in photonic circuits as the electrical diode does in electronic circuits. It could further pave the way to achieve on-chip optical logical devices without nonlinearity or magnetism and bloom the photonic network integration.

Figure 20(a) shows the schematic configuration of the original diode structure under study, which consists of two PC slab domains (PC1 and PC2) with the same lattice constant a but different air hole radii (r_1 and r_2, respectively) comprising a heterojunction structure. These two PC regions stand at a silicon slab [grey area in Fig. 20(a)]. Each PC region has a square-lattice pattern of air holes [white holes in Fig. 20(a)], with the hetero-interface between PC1 and PC2 along the Γ-M direction. We set the two hole radii as a fixed ratio to the lattice constant a, which are $r_1=0.24a$ and $r_2=0.36a$ in order to simplify our discussion. These two composite PCs would comprise a pure PC region if $r_1=r_2$. The light source is placed symmetrically aside the structure with two 4a-wide ridge waveguides connecting the surface of the two PC regions. The whole area is surrounded by a perfectly matched layer.

We simulated the transmission spectra for a TE-like light signal transporting along the forward (from left to right) and backward (from right to left) direction. The refractive index of the dielectric slab was set to 3.4, corresponding to that of silicon at 1,550 nm. The slab thickness was h=0.5a. Figure 20(b)shows the calculated forward (black line) and backward (red line) transmission spectra. The frequency is normalized by a/λ. It is clearly seen that there exists an isolation band ranging from 0.2649 to 0.2958 (a/λ), where the forward transmission forms a peak with a transmissivity of about 6% while the backward transmissivity is down between 0.5% and 1%. The forward peak is located at 0.2793 (a/λ), just in the middle of the isolation band. We define $S=(T_F-T_B)/(T_F+T_B)$ as the signal contrast of the diode, where T_F and T_B denote the forward and backward transmissivity, respectively. The maximum $S$of this original diode equals 0.846 at the peak. Besides, there exists another isolation region from 0.2196 to 0.2649 (a/λ), where the backward transmissivity is higher than the forward transmissivity. This structure thus shows an extraordinary phenomenon of unidirectional transport property.

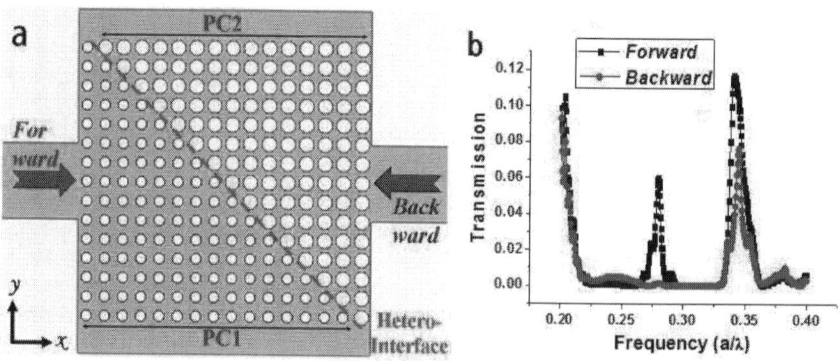

Figure 20. a) Schematic geometry of an original heterojunction optical diode formed by the interface (normal to the Γ-M direction) between two PC slabs (denoted as PC1 and PC2) with different hole radii ($r_1$and r_2, respectively). (b) Simulated transmission spectra of the diode in the forward direction (the black line) and the backward direction (the red line). An input and output ridge waveguide has been used in the 3D-FDTD calculation. [51].

We calculated the band diagram of the TE-like modes of these two PC slabs using the 3D-FDTD method. Figures 21(a) and 21(b) show the calculation results. The first band (even mode) in bulk PC2 [Fig. 2(b)] is directional as the top mode frequency in the Γ-X direction (x-axis) is 0.2345 (a/λ) but that in the Γ-M direction (45°-direction) is 0.3087 (a/λ). Inside the region between 0.2345 (a/λ) and 0.3087 (a/λ), the all-directional transparent region of PC1 needs to be above 0.2633 (a/λ) [Fig. 21(a)] in order to match the bottom mode frequency in the Γ-X direction of the second band (odd mode). These two modes in PCs are the basic working mode of the diode structure. Here the even and odd modes are defined with respect to the off-slab mirror-reflection symmetry σ_z of the field component Ey. In the region between 0.2633 (a/λ) and 0.3087 (a/λ) PC1 is transparent in all directions, while PC2 is transparent along the Γ-M direction but opaque along the Γ-X direction. Compared with Fig. 20(b), the unidirectional transport region [0.2649 to 0.2958 (a/λ)] just coincides with the overlapped region between the directional bandgap of PC2 and the all-directional pass band of PC1. This simple picture indicates that the current unidirectional transport effect involves two ingredients: (I) directional bandgap of PC2 and (II) all-directional pass band of PC1. Noting that the structure does not obey the spatial inversion symmetry alone the propagating direction, the principle of optical isolation can thus be summarized as follows:

1 Forward. When light goes across PC1 as the odd mode and reaches the hetero-junction along the Γ-X direction, it cannot stay in the Γ-X direction in PC2 further because of the Γ-X directional gap. But the hetero-junction is along the Γ-M direction, so light turns to the hetero-junction and diffracts as the even mode at any Γ-M direction into PC2, which passes through PC2 and eventually outputs.

2 Backward. When light goes directly into PC2 as the even mode, it turns to the two Γ-M direction paths which cannot convert to the odd mode of PC1 in the Γ-X direction and eventually leak out so that it does not output.

Based on the numerical analysis of the optical diode, we have fabricated the original diode structure as well as a revised diode structure (with better performance) in silicon, whose SEM pictures are displayed in Fig. 22(a) and (c), respectively. Figures 22(b) and 22(d) show the theoretical and experimental results of the transmission spectra of the two diodes in the forward and backward directions. In Fig. 22(b) the theoretical forward peak of the original diode structure is at 1,575 nm [0.2793 (a/λ)] and the maximum transmissivity is 6%. The experimental forward peak is at 1,556 nm and the maximum transmissivity is 7%. In Fig. 22(d) the experimental forward peak is at 1,534 nm and the maximum transmissivity is 10% for the revised diode, whereas the theoretical forward peak is at 1,552 nm [0.2834 (a/λ)] with a transmissivity of 13%. The measured signal contrast S equals 0.718 (the original structure) and 0.831 (the revised structure) at the peak frequency. Both experimental peaks inFigs. 22(b) and 22(d) have a nearly 20 nm shift and 50 nm broadening against the theoretical simulations, which is probably due to the imperfections in fabrication. The experiment confirms the existence of the unidirectional transport effect in agreement with the theoretical prediction. Due to the arbitrariness of the lattice constant a, we can freely adjust the working frequency to anywhere as desired. This could be more convenient for the design of realistic photonic devices.

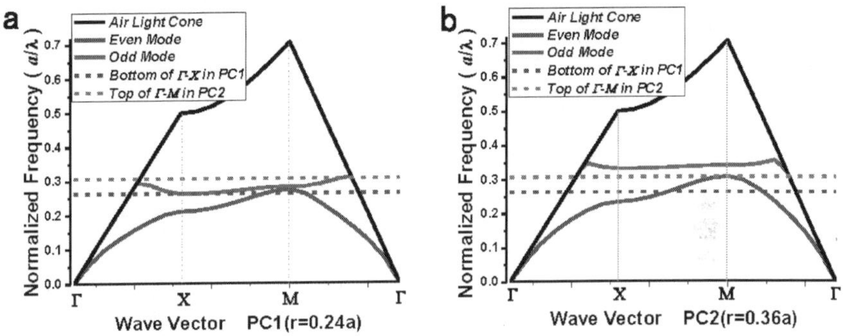

Figure 21. a) Calculated modal dispersion curve for PC1 (r=0.24*a*). (b) Calculated modal dispersion curve for PC2 (*r*=0.36*a*), in which the black line is the air light cone. The red curve denotes the first even mode, while the green curve denotes the second odd mode. The blue dashed line denotes the bottom frequency of the Γ-X directional odd mode [0.2633 (*a*/λ)] of PC1 and the orange dashed line denotes the top frequency of the Γ-M directional even mode of PC2 [0.3087 (*a*/λ)] [51].

The principle for optical diode as analyzed in the above is robust as it is based on a simple directional bandgap mismatch effect of photonic crystal heterojunction. Yet, it should be noticed that in Figs. 22(b) and 22(d) the backward transmissions are fluctuating within 1% to 2% and both are higher than the simulation values, as a result, the signal contrast *S* degrades from 0.846 for the original structure and 0.92 for the revised structure in theory to 0.718 for the original structure and 0.831 for the revised structure in experiment. The performance improvement of the diode relies on how to maximize the peak of the forward transmissivity and minimize the backward transmissivity in experiment. Several means can help improve the forward transmissivity. First, one can change the air hole size of PC1 and PC2 and enlarge the directional bandgap. Calculations show that the forward peak transmissivity of the revised structure with r_1 =0.30*a* and r_2 =0.45*a* grows up dramatically to 29.4% while maintaining the same low level of backward transmissivity. Second, one can change the relative size of the input and output waveguides. Calculations show that by changing the input waveguide width to 2*a* and keeping the output waveguide width 4*a*, the forward peak signal increases up to 20.8% in transmissivity.

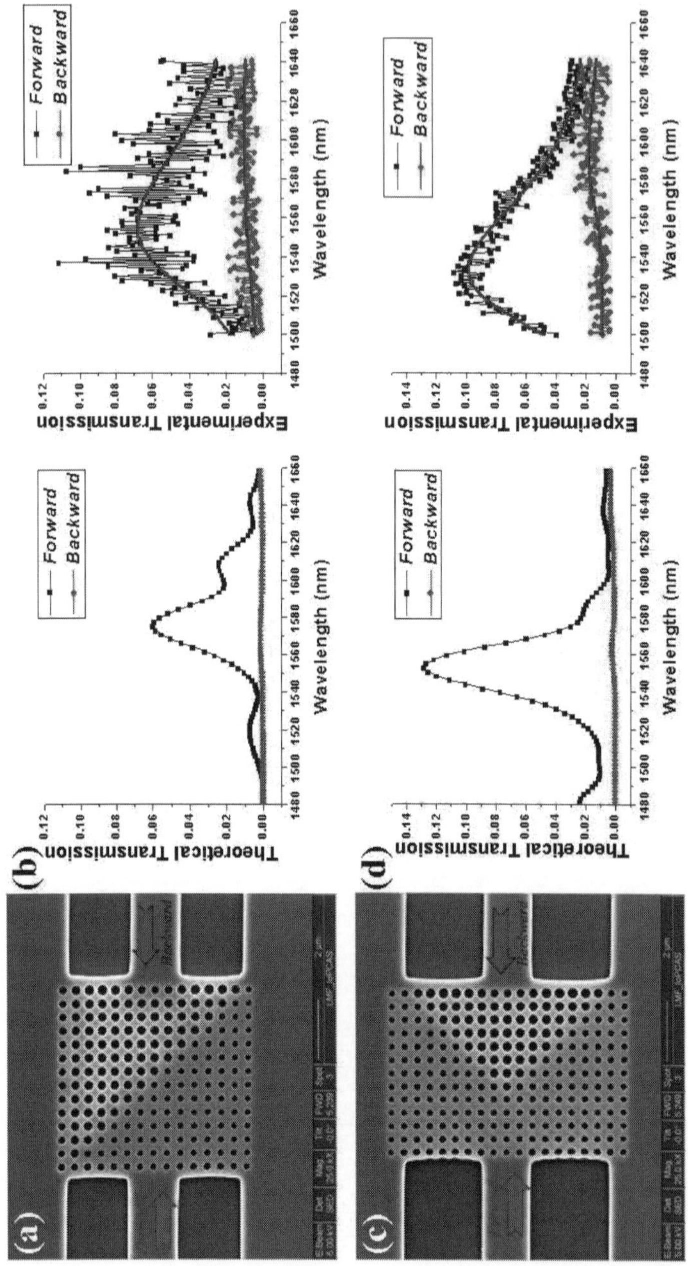

Figure 22. a) Scanning electron microscope images of the original optical diode structures. (b) Theoretical (left)and experimental (right) transmission spectra of the original diode structure. (c) Scanning electron microscope images of the revised optical diode structures. (d) Theoretical (left) and experimental (right) transmission spectra of the revised diode structure [51].

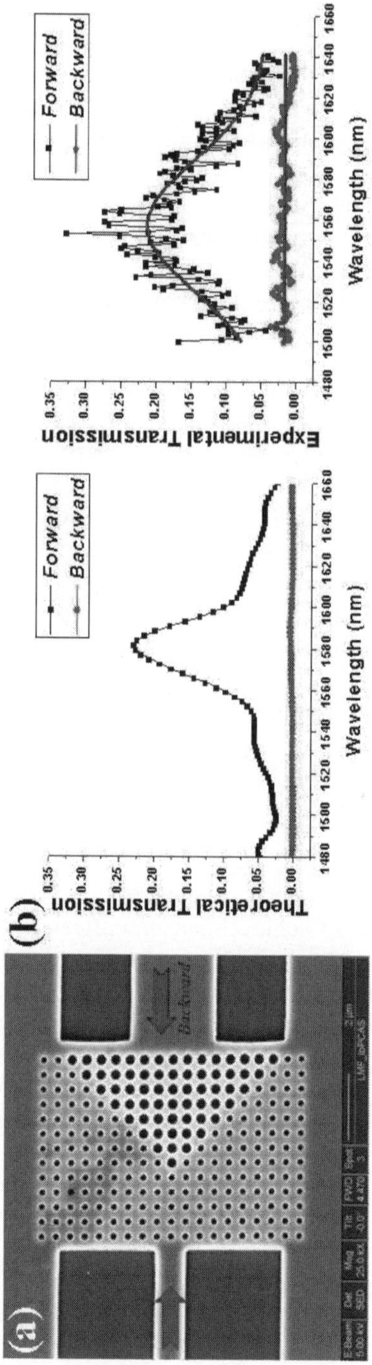

Figure 23. a) Scanning electron microscope images of the optimized optical diode system. (b) Theoretical (left) and experimental (right) transmission spectra of the optimized diode structure [51].

To reduce the backward transmissivity, one can either enlarge the directional bandgap of PC2 to attenuate the backward signal more strongly, or eliminate the return of leak-out light from the outside of slab or the structure boundary by introducing the absorbing metal dots near the structure, or enlarge the heterojunction structure appropriately so that the leak-out light cannot enter the output waveguide. Following the above general ideas, we further optimize the optical diode structures as illustrated in Fig. 23. The structure has parameters of r_1=100 nm and r_2=160 nm, and the input and output waveguide width are $2a$ and $6a$, respectively. In experiment, we have got an optical diode with an maximum of 32.8% of the forward peak transmissivity and 0.885 of the signal contrast S at 1,557 nm.

Our PC heterojunction diode has advantages of high signal contrast, wavelength-scale small sizes, and being all-dielectric, linear, and passive. Furthermore, it has a much smaller scale than those based on diffraction gratings and thus greatly facilitates large-scale integration. The high performance on-chip optical diode realized in silicon without nonlinearity or magnetism will stimulate the exploration of other more complex on-chip optical logical devices with ultra-high stability, integration and much less power consumption. Such an optical diode may play the same basic role in photonic circuits as the electrical diodes do in electronic circuits, which have significantly revolutionized fundamental science and advanced technology in various aspects of our routine life due to their capability of rectification of current flux. Furthermore, its large-scale fabrication could be readily achieved by the well-developed CMOS techniques. The realization of high-performance on-chip optical diodes may open up a road toward photonic computers.

Strictly speaking, the existence of unidirectional transport effect of light does not mean automatic achievement of optical isolation. Recently there have appeared hot controversies upon whether isolation of light can be realized via linear and passive photonic structures. Several schemes to realize unidirectional transport of light through linear and passive photonic structures have been proposed, which are essentially based on the principle of spatial-inversion symmetry breaking. Feng *et al.* reports a passive silicon optical diode based on one-way guided mode conversion [52]. However, whether or not nonreciprocal transport of light can happen in the structure has raised hot controversies [53,54]. Fan*et al.* made a scattering matrix analysis for relevant forward and backward modes of the structure and argued that the structure is essentially reciprocal and cannot enable optical isolation because it possesses a symmetric coupling scattering matrix. In their response, Feng *et al.* acknowledge that their structure, as a one-way mode converter with asymmetric mode conversion, is Lorentz reciprocal, which states that the relationship between an oscillating current and the resulting electric field is unchanged if one interchanges the points where the current is placed and where the field is measured, and on its own cannot be used as the basis of an optical isolator. The controversies have thus raised a fundamental question: Can one construct an optical isolator by using a linear and time-independent optical system? The answer to this question by the authors of Ref. [53,54] obviously is no.

But our theoretical and experimental study on the optical isolation performance of our PC heterojunction diode leads to a totally different answer to the above question, namely, the spatial inversion symmetry breaking diode can construct an optical isolator in no conflict with any reciprocal principle [55]. To see whether there is a good isolation effect of the silicon diode, we implement a direct method that is originated from the conventional magneto-optical isolator that has been popularly used in laser devices [Fig. 24(a)]. One places a total reflection mirror after the output port in the forward direction of the isolator device and monitor the reflection signal from the input port. This reflection signal well describes and measures the round-trip transmissivity of light across the isolator device. If the reflection signal is the same as or is comparable with the forward signal, then the structure does not have the desired isolation property. In contrast, if the reflection signal is much smaller than the forward signal, then a good isolation property is implied.

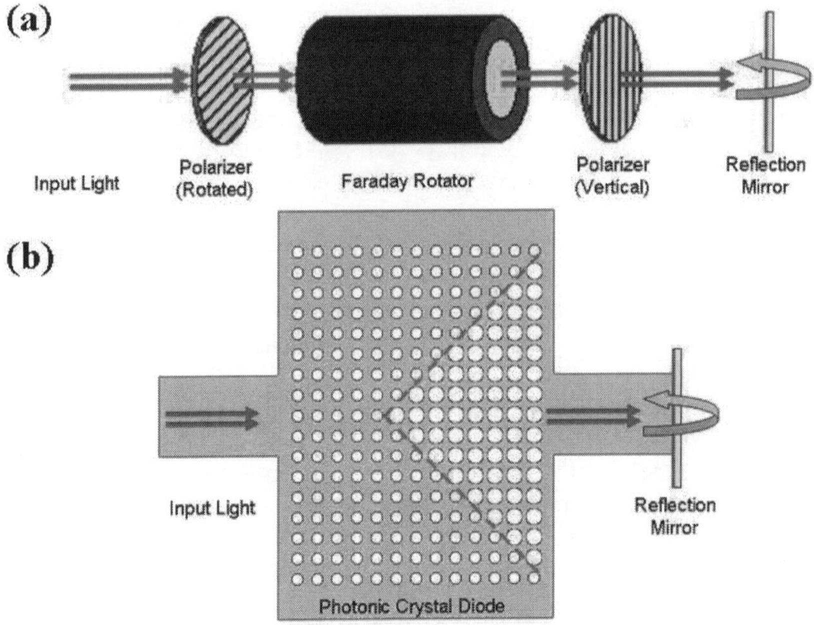

Figure 24. a) Traditional magnetic-optical isolator with reflection mirror at the output port. (b) On-chip optical diode system with reflection mirror at the output port. The absence of the reflection signal at the input port can indicate the isolation property of the system [55].

An equivalent way to investigate the optical isolation performance of the diode structure is to adopt a doubled-diode structure with a mirror-symmetrical plane at the forward direction output port of the diode, as depicted in Fig. 25. Obviously this method has set all the forward output signals as the backward input signal of the diode and thus can directly test the isolation property of the diode structure. By implementing this method, we calculate simultaneously the forward transmissivity and the round-trip transmissivity of the diode structure. Comparison of these two quantities would directly measure their isolation properties. Figure 25(a) is the schematic geometry of the doubled-diode structure corresponding to the diode depicted in Fig. 24(b). The parameters of the diode are the same as in Fig. 23. The width of the input and round-trip output waveguides is the same $2a$ (a=440 nm) and the center connection waveguide is $6a$. The length of the center connection waveguide, which measures the distance of the total-reflection mirror away from the output port of the diode, is $10a$. The spectra [Fig. 25(b)] show that the round-trip reflection peak is located at 1,582 nm and has a quantity of only 0.3%, which is almost two orders of magnitude smaller than the forward peak [with a maximum transmissivity of 22.9%]. The result indicates that the diode has a significant optical isolation property.

Based on the above numerical analysis, the double-diode structure was fabricated in silicon. Figure 25(c) shows the SEM images of the fabricated double-diode structures along the light path. The lattice constant a was 440 nm, and the radii r_1 and r_2 of air holes in the two photonic crystals of the heterojunction diode were approximately 110 nm and 160 nm. The length of the output waveguide is 4 μm (≈$10a$). The slab thickness was 220 nm. Figure 25(d) shows the experimentally measured forward and backward transmission spectra of the diode, as well as the round-trip transmission spectrum of the doubled-diode structure. The forward, backward and round-trip transmissions are optimized and the input/output loss has been removed. The spectra show that the maximum round-trip transmissivity, located at 1,553 nm, is only 0.4%, almost two orders of magnitude smaller than the forward peak [with a maximum transmissivity of 32.9%]. The experiment confirms the existence of a significant isolation effect in agreement with the theoretical prediction. We change the length of the center connection waveguide of the double-diode structure from 4 μm to 3 μm and 5 μm. The measured round-trip transmission spectra for the three structures are displayed in Fig. 25(e). The results show that the round-trip transmission signal decreases remarkably along with the increasing length of the center connection waveguide, and already reaches an extremely low level (below 0.05%) in the whole spectrum range when the output waveguide length increases to 5 μm. This clearly indicates that the mode dispersion in the output waveguide of the diode has no influence to the isolation property of the diode. Due to the arbitrariness of the lattice constant a, we can freely adjust the isolation frequency to anywhere as desired. This could be very convenient for the design of realistic photonic devices.

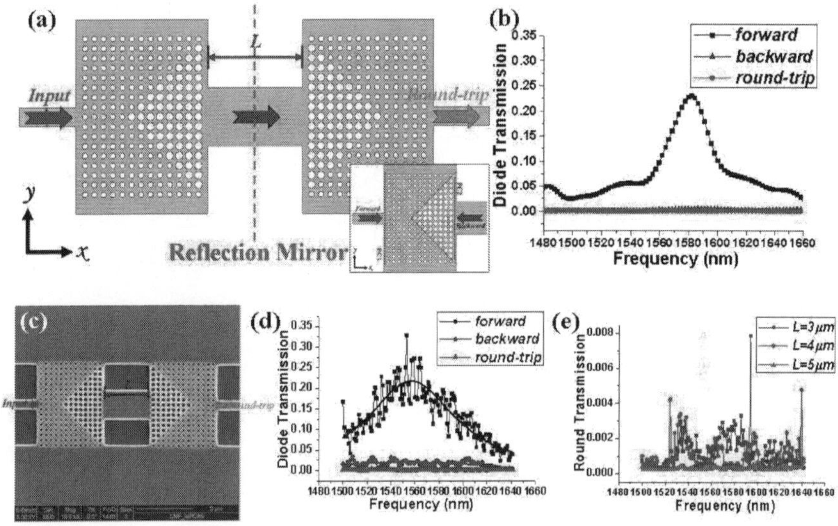

Figure 25. a) Schematic geometry of a doubled-diode structure with the total reflection mirror modeling the round-trip transmission of an isolator, and the corresponding single-diode structure under forward and backward transmissions. (b) Calculated forward (black line), backward (blue line), and round-trip (red line) transmission spectra of the diode. (c) Scanning electron microscope images of doubled-diode structure. (d) Experimental transmission spectra of forward transmission (black line), backward transmission (blue line), and the round-trip reflection (red line). (e) Experimental spectra of the round-trip reflection with changed center waveguide length L [55].

To better understand the underlying physics, we further make a detailed analyses based on the scattering matrix theory adopted in Ref. [53,54], and find that the above numerical results of optical isolation are in no conflict with the reciprocity theorem involved in our linear and passive silicon optical diode structure. Our diode basically consists of two in-plane information channels (A and B, the input and output waveguide channels for infrared signal, which can be either single mode or multimode channels.) as well as many in-plane side-way and off-plane scattering channels (denoted as C as a whole, which causes dissipation of information away the signal channels). At the two ends of the diode device the fields are written as follows:

$$\begin{bmatrix} A_{out} \\ B_{out} \\ C_{out} \end{bmatrix} = S \begin{bmatrix} A_{in} \\ B_{in} \\ C_{in} \end{bmatrix}, \tag{3}$$

in which A_{in} corresponds to the input signal from port A, A_{out} to the output signal from port A, B_{in} to the input signal from port B, B_{out} to the output signal from port B, C_{in} to the input signal from port C, and C_{out} to the output signal from port C. The scattering matrix S transforms the input state of all the channels [the column vector in the right hand of Eq. (3)] into the output state of all the channels [the column vector in the left hand of Eq. (3)]. The scattering equation of the forward transmission is written as:

$$\begin{bmatrix} a_{out1} \\ b_{out1} \\ c_{out1} \end{bmatrix} = S \begin{bmatrix} a_{in0} \\ 0 \\ 0 \end{bmatrix}. \tag{4}$$

As the silicon diode structure is linear and passive, the system as a whole is reciprocal in regard to time-reversal symmetry following the Lorentz reciprocity theorem. As a result, the scattering matrix S is symmetric with $S=S^T$ and further satisfies:

$$S^* = S^{-1}. \tag{5}$$

Suppose all the output signals are reversed and come back into the system, then the input at port B for the system is now exactly the same as $[a^*_{out1} \;\; b^*_{out1} \;\; c^*_{out1}]^T$. The scattering equation is then

$$S\begin{bmatrix} a^*_{out1} \\ b^*_{out1} \\ c^*_{out1} \end{bmatrix} = (S^* \begin{bmatrix} a_{out1} \\ b_{out1} \\ c_{out1} \end{bmatrix})^* = (S^{-1} \begin{bmatrix} a_{out1} \\ b_{out1} \\ c_{out1} \end{bmatrix})^* = \begin{bmatrix} a^*_{in0} \\ 0 \\ 0 \end{bmatrix}, \tag{6}$$

which is exactly the same as the initial input from port A. This clearly indicates that there is no isolation behavior in the structure if all information is reversed back into the system, consistent with the reciprocity theorem for a time-reversal symmetric system.

However, the story can be very different when the in-plane signal transport is concerned, as is always the case for 2D silicon PC slab structures. In our structure the information and energy involved in C channels are dissipated permanently against the in-plane channel A and B due to scattering loss (both in-plane and off-plane), and they cannot be reversed back totally and input again into the structure, so in practice, C_{in} in Eq. (3) can be assumed to be zero. As a result, Eq. (6) should be modified as:

$$S\begin{bmatrix}0\\ b^*_{out1}\\ 0\end{bmatrix}=\begin{bmatrix}a_{out2}\\ b_{out2}\\ c_{out2}\end{bmatrix}. \tag{7}$$

In general, Eq. (7) looks very different from Eq. (6), which indicates that the reciprocal transport of light in regard to the signal channel A and B has been broken. It shows that even if the same forward transmission signal of port B is reversed back and input into the diode, the output signal of port A can be much different from the initial input signal a_{in0} of port A because no signal is reversed and input back into the channel C. Therefore, the considerable unidirectional transmission behavior can take place for the in-plane signal with no conflict with the reciprocal principle. In other words, the $a_{out2}(=S_{12}\bullet b^*_{out1})$ could be much different from $a^*_{in0}(=S_{11}\bullet a^*_{out1}+S_{12}\bullet b^*_{out1}+S_{13}\bullet c^*_{out1})$ when $S_{11}\bullet a^*_{out1}+S_{13}\bullet c^*_{out1}\neq 0$. This justifies the occurrence of a good isolation effect in the silicon optical diode. In ideal structures, both of them are zero, and Eq. (7) becomes

$$S\begin{bmatrix}0\\ b^*_{out1}\\ 0\end{bmatrix}=\begin{bmatrix}0\\ 0\\ c_{out2}\end{bmatrix}, \tag{8}$$

which implies a 100% signal contrast of the isolator.

It is worth saying a few more words here for better drawing a clear picture about the physics discussed in the above. In nature, as time always flows forward and cannot be reversed, one usually uses the term of reciprocal or nonreciprocal transport of light to describe a model system of back transport of light, in many cases to describe the reflection of light back into the considered structure. In this regard, simply consider a point source radiating an outgoing spherical wave front. If time can be reversed, the outgoing spherical wave front is contracted into an ingoing spherical wave front, eventually to a point. This is a very good picture to describe reciprocal transport of light in a linear system. However, to realize in real world such a concept, one needs to place a perfect spherical mirror concentric with the point source of light, which reflects back all information carried by the outgoing expanding spherical wave into the ingoing contracting spherical wave. If, however, one has only a small planar mirror placed at some distance and with a limited solid angle with respect to the source, the reflected signal can never return to the initial state of a point source when it reaches the position where the light source is located. The conventional magneto-optical isolator also works in this category of physical picture. It is used to block down the back-reflection signal of the transmission light, and the underlying physics can be well described by the model of time-reversal symmetry breaking. The same physics picture applies equally well to our optical diode. The fact that there exists information dissipation from the signal channels to other channels in a spatial-inversion symmetry breaking structure is sufficient to induce an optical isolation in regard to the signal.

The above numerical calculations and experimental results have shown that our silicon PC slab heterojunction diode exhibits promising performance of optical isolation, with a round-trip transmissivity two orders of magnitude smaller than the forward transmissivity for in-plane infrared light across the structure. Our scattering matrix analysis indicates that the considerable unidirectional transport of in-plane signal light can be attributed to the information dissipation and selective modal conversion in the multiple-channel spatial-inversion symmetry breaking structure and has no conflict with reciprocal principle for a time-reversal symmetric structure. It is expected that optimized connection interfaces between the input and output waveguides with the heterojunction diode can yield better impedance mismatch and bring higher forward transmission efficiency. That optical isolation can occur in a linear, passive, and time-independent optical structure would stimulate more thinkings and insights on the general transport theory of light in the fundamental side and open up a road towards photonic logics in silicon integrated optical devices and circuits in the application side.

7. PARALLEL–HETERO PHOTONIC CRYSTAL STRUCTURES

Photonic crystal heterostructures (PCHs) have attracted increasing interest in optical integrated circuits and cavity quantum electrodynamics (cavity QED) due to their useful photonic band-gap structure and the ability to provide nanocavities with ultra-high quality factor (Q factor). The properties of

PCHs have been investigated both in theory and experiment over the past several years. In previous works, the transmission and reflection characteristics of PCHs were revealed only across the hetero-interface between two photonic crystals with different lattice constants. The basic character of PCHs was the shift of the band edge, which results in a transmission gap with approximately 100% efficiency [56]. In comparison, recently we have reported a method for making a parallel-hetero perturbation inside the waveguide and analyze the optical properties of the photonic crystal parallel-hetero perturbation (PHP) structure. It is expected that this new type of PCHs not only contains the band-edge shifting property but also has an additional transmission gap which can be easily regulated in the middle of the transmission band. Simulations and experiments confirm the existence of the additional transmission gap [57]. Our work can help to enlarge the usage of PCHs in the design and fabrication of novel cavities and filters via localized modulation of structural dispersion, which are key components in a photonic network. Based on the PCH structures, we have further proposed and realized a new scheme of cavity, an interface heterostructure cavity without any confinement barrier to confine light. Interestingly, the localized resonant mode lies in the pass band of the waveguide, in comparison of those cavities whose localized modes are always located within the band gap of the structures.

Figure 26(a) shows the geometry of the designed cavity, which consists of two identical parallel-hetero perturbation (PHP) waveguides. Each PHP waveguide consists of two semi-infinite PC (PC1 and PC2) slabs having slightly different lattice constants (a_1 and a_2, respectively). We assume a_2 is 5% larger than a_1. Each PC region has a triangular-lattice pattern of air holes, with the W1 waveguide formed by a line of missing holes along the Γ-K direction. Without periodicity along any direction, the PHP is essentially a kind of incommensurate PC superlattice along the waveguide direction. The transmission spectra of PHP waveguide structure exhibits an additional gap [0.29-0.30 (a_1/λ)] in the middle of the pass band of the two individual W1 waveguides besides the usual band edge shifting [0.26-0.27 (a_1/λ)] [56]. Figure 26(b) shows the calculated resonance spectrum of the cavity by using 3D FDTD method in association with the Pade approximation for the TE-like modes of the PC slab [58]. The resonance spectrum shows that a resonant mode surprisingly appears in the pass band region of the two PHP waveguides at 0.2855 (a_1/λ). In addition, this resonant mode (called band-pass mode) has a rather high value of Q factor reaching 5,340.

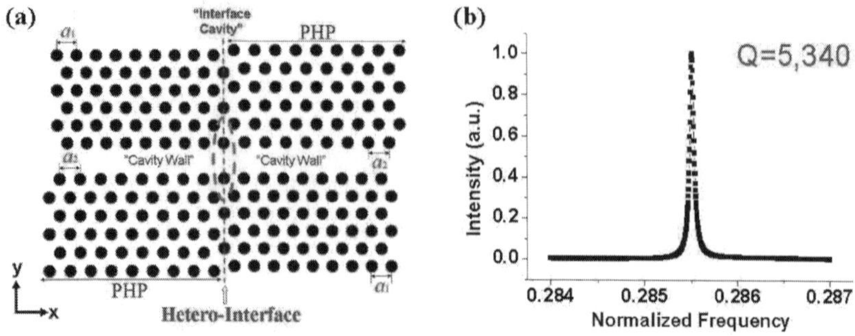

Figure 26. a) Schematic geometry of an anti-symmetric parallel-hetero cavity structure formed by only two PHP waveguides. (b) Calculated resonant mode distribution of the interface PHC [58].

We calculated the E_y field distribution of TE-like mode transporting along the waveguide at the resonance frequency 0.2855 (a_1/λ). It is surprising that the interface PHC has no influence on the propagation of waveguide mode [Fig. 27(a)]. To further confirm that the resonant mode really exists, we plot in Fig. 27(b) the calculated Ey field distribution of the resonant mode at the interface PHC at 0.2855 (a_1/λ) using a point source located within the interface cavity region. The mode does oscillate as a quadrupole form without barriers along the propagation direction even after the light source is turned off. This indicates that the resonant mode does exist and is not a calculation fault.

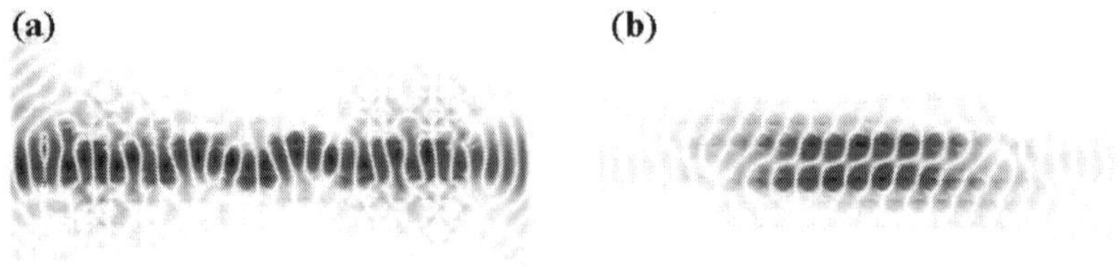

Figure 27. a) Calculated TE-like mode Ey field distribution of light transporting along the waveguide at the resonance frequency 0.2855 (a_1/λ). (b) Calculated TE-like mode Ey field distribution of the interface-cavity resonance at the resonance frequency 0.2855 (a_1/λ) [58].

Based on the numerical analysis, an interface PHC structure was fabricated [Fig. 28(a)].The experimental results [Fig. 28(b)] confirms the theoretical prediction with the Q factor of the cavity resonant mode decreases to 1,517 (centered at 1,570.2 nm) [Fig. 28(c)]. We then use an infrared CCD camera to monitor the resonant properties on top of the interface PHC region. When the laser wavelength is tuned to 1,570 nm, we can readily see a bright light spot located at the exact interface position [red circle in Fig. 28(d)] besides the other two light spots induced by the coupling between PHP waveguide and ridge waveguide. When the laser wavelength is tuned to 1,560 nm, the light spot in the red circle is gone while the other two light spots remain, indicating that the cavity is off resonance. This confirms the existence of the band-pass resonant mode.

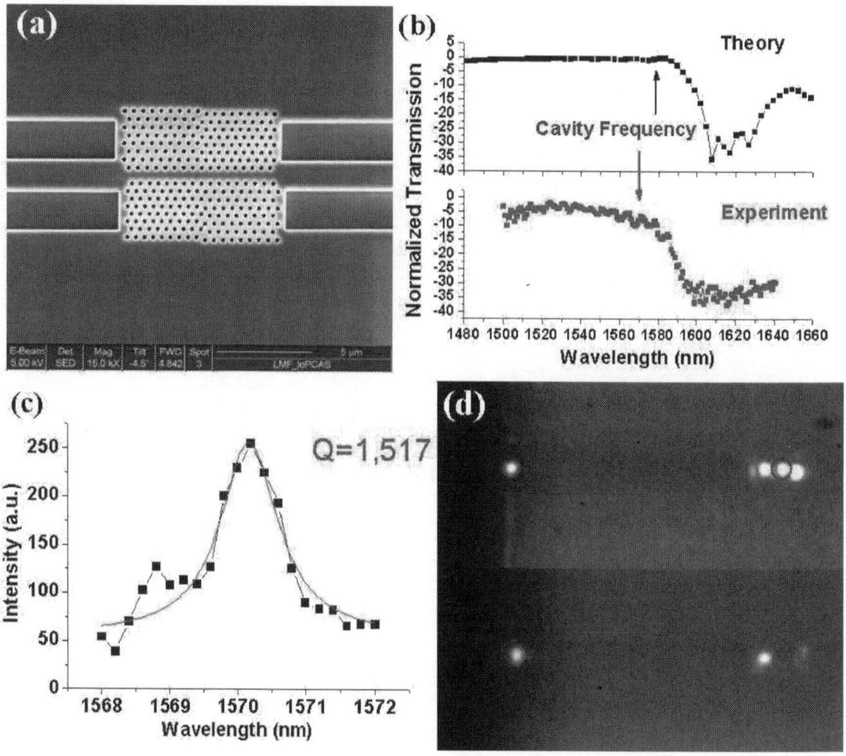

Figure 28. a) SEM image of the fabricated interface PHC composed of two PHP waveguides. (b) Experimental transmission spectrum of the interface PHC compared with the simulation results. (c) Resonance spectrum of the light spot obtained by its gray class analysis. (d) Infrared CCD images of the interface PHC structure on resonance (1,570 nm) and off resonance (1,560 nm) with output ridge waveguide [58].

The physics of PHC can be understood from the point of view of slow light effect. We have used a narrow light pulse centered at 0.2855 (a_1/λ) to pass through the interface PHC and a pure, same-length PHP waveguide, respectively, and recorded the output time of the pulse. The interface PHC is 599.25 time units, while the pure PHP waveguide is 606.3 time units, slightly longer than the interface PHC [Fig. 29(a)]. This indicates that the interface still contains one PC period length, so the whole interface PHC can be seen as a three-part structure, a combination of "slow light"–"fast light"–"slow light" region [Fig. 29(b)]. The central fast region in the PHC, which is only one period long, could also confine energy between two slow light regions and behaves as a cavity.

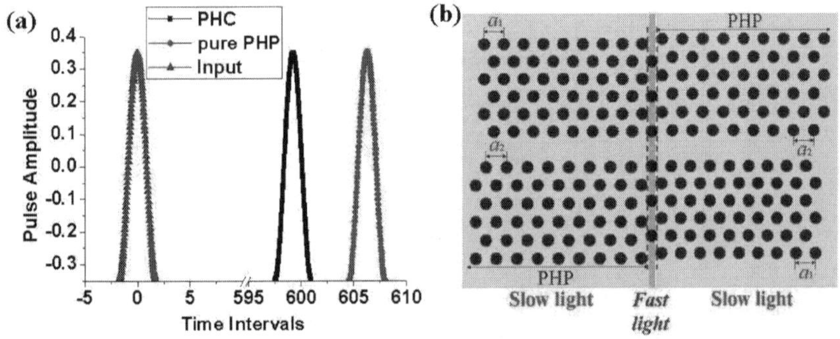

Figure 29. a) Time diagram of a narrow light pulse at 0.2855 (a_1/λ) passing through the interface PHC and a pure, same-length PHP waveguide. (b) Schematic slow light diagram of interface PHC [58].

The physics underlying the interface PHC can also be analyzed by focusing on the phase shift in the waveguide of the interface PHC during propagation due to the fact that this band-pass mode is near the edge of the additional-gap region. The current PHC, which is made from an incommensurate superlattice structure, is similar to a periodic structure with dislocations. As a result, multiple scattering of light around the dislocation will occur and result in pinning effect of light (energy shifting across the dislocation line). Previous study shows that the PHP structure provides the asymmetric phase condition during the light propagation along the waveguide. The interface connects two same anti-symmetric PHPs and can be seen as a vertical edge dislocation line (phase reversal line), so that the phase condition is reversed to the opposite aside the interface. In material science the dislocation could cause the charge accumulation. Similarly our phase reversal dislocation gives an abrupt phase shift across the interface and may cause the photon accumulation around the phase reversal line. This phase reversal dislocation plays as a defect and forces the light energy distribution to become skew and swinging while crossing the interface [Fig. 27(a)]. The local energy oscillation acts as the cavity resonance and leads to the photon accumulation at the interface.

The PHC has several distinct properties. First, unlike the conventional PC cavities which are formed either by removing one or several air holes or by waveguide-like hetero-cavities, the current PHC does not have any confinement barrier to confine light. Second, the resonant mode is located within the pass band of waveguides, so the localized resonant mode and the continuous waveguide mode can easily co-exist in the same frequency and space regime. Third, the coupling efficiency between waveguide and cavity modes is much higher than the conventional cavities, which strongly depend on the transverse evanescent field profile overlap of the cavity mode and the waveguide mode. As a result, this PHC has nearly no influence on the propagation of waveguide mode despite the presence of localized cavity mode. High coupling efficiency between waveguide and cavity mode can help to reduce the operating power for PC lasers, and make it possible to integrate multiple lasers, photodetectors, and switches on a single chip.

8. CONCLUSIONS

In summary, we have presented recent progresses on infrared 2D air-bridged silicon PC slab devices that were made in our group in the past several years. 2D air-bridged silicon PC slab structures can confine light by the high index contrast in the vertical off-plane direction and manipulate light by photonic band and band gaps in the lateral in-plane direction. In addition, silicon is transparent and has a large refractive index in the infrared wavelengths. Therefore, the air-bridged silicon PC structures become one of the most important elements in integrated optics.

In this review, we have discussed several integrated optical elements and devices in regard to their design, fabrication, and characterization. These devices are based on either the PBG or photonic band structure engineering. To bring these devices into reality, we have made extensive efforts to construct high-efficiency numerical simulation tools for solution of photonic band structures, transmission spectra, light propagation dynamics, light wave patterns, and many others. These simulations allow for design of optimized PC structures for a specific application. The fabrication of these devices strongly depends on state-of-the-art nanofabrication technologies including EBL, FIB, ICP etching, and other wet etching techniques and procedures. We have constructed a convenient experimental setup to measure the transmission spectra and monitor the propagation path of infrared signals within PC structures simultaneously. The setup makes the optical characterization of PC devices accurate, user-friendly, fast, and convenient.

We have proposed Γ−M waveguides made in 2D triangular-lattice PC slabs. After a series of optimization and improvement, we find out the optimized geometries for high performance wide band Γ−M waveguides. As the Γ−M waveguide is perpendicular to the usual Γ−K waveguide, we combine the Γ−K waveguide and Γ−M waveguide together to form 90° waveguide bends and channel drop filters. The combination of Γ−M waveguide and Γ−K waveguide can offer a flexible way to interconnect and couple between different devices. In addition, we have shown design of two other kinds of channel drop filters. They achieve fine tuning of the resonant wavelengths by changing the size of the cavities or the shape of the air holes around the cavities. We have also designed and realized PC waveguide based on the coupled-cavity scheme. There are a lot of geometric parameters to fine tune the light transport properties in the structure. Moreover, we have successfully fabricated high-Q silicon PC microcavities with a Q-factor up to 70000 and this paves the way to experimentally explore light-matter interaction within the strong-coupling regime in the all-solid platform.

We have investigated optical devices that work on band structure engineering. We have explored PC structures that exhibit interesting and useful dispersion and refraction properties, such as negative refraction and self-collimation effects. We have designed and fabricated a kind of PC structure with negative refraction effect and use SNOM technology to observe the negative refraction ray trace of infrared light beam. In addition, we obtain broadband large-angle self-collimation effect for TE-like guided modes in infrared wavelength.

We have demonstrated the design, fabrication, and characterization of on-chip wavelength-scale optical diodes that are made from the heterojunction between two different silicon 2D square-lattice PC slabs with directional bandgap mismatch and different mode transitions. The measured transmission spectra show considerable unidirectional transmission behavior, in good agreement with numerical simulations. The experimental realization of on-chip optical diodes with wavelength-scale size using all-dielectric, passive, and linear silicon photonic crystal structures may help to construct on-chip optical logical devices without nonlinearity or magnetism, and would open up a road towards photonic computers.

We have demonstrated optical isolation of our diode structure. Both numerical simulations and experimental measurements show that the round-trip transmissivity of our diode could be two orders of magnitudes smaller than the forward transmissivity, indicating good performance of optical isolation. The occurrence of in-plane light isolation is attributed to the information dissipation due to off-plane and side-way scattering and selective modal conversion in the multiple-channel structure and has no conflict with the reciprocal principle. That optical isolation can occur in a linear, passive, and time-independent optical structure would stimulate more thinkings and insights on the general transport theory of light in the fundamental side and open up a road towards photonic logics in silicon integrated optical devices and circuits in the application side.

We have designed and fabricated cavities without confinement barrier by combining two incommensurate PC superlattice waveguides. A resonant mode with a high quality factor shows up in the pass band of waveguides. It has nearly no influence on the propagation of waveguide mode and can be directly coupled with the waveguide mode. The experimental measurement confirms the theoretical prediction of extraordinary coexistence of localized cavity mode and continuous waveguide mode with high coupling efficiency in the same frequency and space regime. The novel type of cavity has a number of unique properties that are advantageous to on-chip information transport and processing. The discovery of cavity without confinement barrier might attract interest in fundamental physics and optical engineering communities.

All of these results show that 2D air-bridged silicon PC structures can control light propagation in many flexible ways and have many potential applications in all-optical integrated circuits and other fields. The efforts that we have made and the experiences that we have accumulated in the past several years will allow us to design and realize PC devices with novel functionalities.

9. ACKNOWLEDGEMENTS

This work was supported by the National Key Basic Research Special Foundation of China (No. 2011CB922002), the National Natural Science Foundation of China, and National Center for Nanoscience and Technology of China. The authors would like to thank the previous members of our group, Cheng Ren, Haihua Tao, Yazhao Liu, and Changzhu Zhou for their contributions to the works presented in this review.

REFERENCES

1 E Yablonovitch, Inhibited spontaneous emission in solid-state physics and electronics. Physical Review Letters, 1987

2 J D Joannopoulos, S G Johnson, J N Winn, R D Meade, Photonic Crystals: Molding the Flow of Light. 2nd ed. Princeton: Princeton University Press, 2008

3 J Valentine, S Zhang, T Zentgraf, E Ulin-avila, D A Genov, G Bartal, X Zhang, Three-dimensional optical metamaterial with a negative refractive index. Nature, 2008

4 J Yao, Z W Liu, Y M Liu, Y Wang, C Sun, G Bartal, A M Stacy, X Zhang, Optical negative refraction in bulk metamaterials of nanowires. Science, 2008

5 S G Johnson, S H Fan, P R Villeneuve, J D Joannopoulos, L A Kolodziejski, Guided modes in photonic crystal slabs. Physical Review B: Condensed Matter and Materials Physics, 1999

6 O Painter, R K Lee, A Scherer, A Yariv, O Brien, J D Dapkus, P D Kim, I , Two-dimensional photonic band-gap defect mode laser. Science, 1999

7 S J Mcnab, N Moll, Y A Vlasov, Ultra-low loss photonic integrated circuit with membrane-type photonic crystal waveguides. Optics Express, 2003

8 C Y Luo, S G Johnson, J D Joannopoulos, J B Pendry, All-angle negative refraction without negative effective index. Physical Review B: Condensed Matter and Materials Physics, 2002

9 H Kosaka, T Kawashima, A Tomita, M Notomi, T Tamamura, T Sato, S Kawakami, Self-collimating phenomena in photonic crystals. Applied Physics Letters, 1999

10 X F Yu, S H Fan, Bends and splitters for self-collimated beams in photonic crystals. Applied Physics Letters, 2003

11 H Kosaka, T Kawashima, A Tomita, M Notomi, T Tamamura, T Sato, S Kawakami, Superprism phenomena in photonic crystals: toward microscale lightwave circuits. Journal of Lightwave Technology, 1999

12 T Baba, T Matsumoto, M Echizen, Finite difference time domain study of high efficiency photonic crystal superprisms. Optics Express, 2004

13 A Berrier, M Mulot, M Swillo, M Qiu, L Thylén, A Talneau, S Anand, Negative refraction at infrared wavelengths in a two-dimensional photonic crystal. Physical Review Letters, 2004

14 M Born, E Wolf, A B Bhatia, Principles of Optics: Electromagnetic Theory of Propagation, Interference and Diffraction of Light. New York : Cambridge University Press, 1999

15 K M Ho, C T Chan, C M Soukoulis, Existence of a photonic gap in periodic dielectric structures. Physical Review Letters, 1990

16 S Johnson, J Joannopoulos, Block-iterative frequency-domain methods for Maxwell's equations in a planewave basis. Optics Express, 2001

17 Z Y Li, B Y Gu, G Z Yang, Large absolute band gap in 2D anisotropic photonic crystals. Physical Review Letters, 1998

18 Z Y Li, J Wang, B Y Gu, Creation of partial band gaps in anisotropic photonic-band-gap structures. Physical Review B: Condensed Matter and Materials Physics, 1998

19 J B Pendry, Photonic Band Structures. Journal of Modern Optics, 1994

20 C T Chan, Q L Yu, K M Ho, Order-N spectral method for electromagnetic waves. Physical Review B: Condensed Matter and Materials Physics, 1995

21 A Taflove, Computational Electrodynamics: the Finite-Difference Time-Domain Method. Boston : Artech House, 1995

22 N A Nicorovici, R C Mcphedran, L C Botten, Photonic band gaps for arrays of perfectly conducting cylinders. Physical Review E: Statistical Physics, Plasmas, Fluids, and Related Interdisciplinary Topics, 1995

23 L M Li, Z Q Zhang, Muitiple-scattering approach to finite-sized photonic band-gap materials. Physical Review B: Condensed Matter and Materials Physics, 1998

24 Z Y Li, L L Lin, Photonic band structures solved by a plane-wave-based transfer-matrix method. Physical Review E: Statistical, Nonlinear, and Soft Matter Physics, 2003

25 Z Y Li, L L Lin, Evaluation of lensing in photonic crystal slabs exhibiting negative refraction. Physical Review B: Condensed Matter and Materials Physics, 2003

26 L L Lin, Z Y Li, K M Ho, Lattice symmetry applied in transfer-matrix methods for photonic crystals. Journal of Applied Physics, 2003

27 Z Y Li, K M Ho, Light propagation in semi-infinite photonic crystals and related waveguide structures. Physical Review B: Condensed Matter and Materials Physics, 2003

28 Z Y Li, L L Lin, K M Ho, Light coupling with multimode photonic crystal waveguides. Applied Physics Letters, 2004

29 M Che, Z Y Li, Analysis of photonic crystal waveguide bends by a plane-wave transfer-matrix method. Physical Review B: Condensed Matter and Materials Physics, 2008

30 M Che, Z Y Li, Analysis of surface modes in photonic crystals by a plane-wave transfer-matrix method. Journal of the Optical Society of America a-Optics Image Science and Vision, 2008

31 Z Y Li, K M Ho, Analytic modal solution to light propagation through layer-by-layer metallic photonic crystals. Physical Review B: Condensed Matter and Materials Physics, 2003

32 J J Li, Z Y Li, D Z Zhang, Second harmonic generation in one-dimensional nonlinear photonic crystals solved by the transfer matrix method. Physical Review E: Statistical, Nonlinear, and Soft Matter Physics, 2007

33 Z Y Li, J J Li, D Z Zhang, Nonlinear frequency conversion in two-dimensional nonlinear photonic crystals solved by a plane-wave-based transfer-matrix method. Physical Review B: Condensed Matter and Materials Physics, 2008

34 A F Oskooi, D Roundy, M Ibanescu, P Bermel, J D Joannopoulos, S G Johnson, MEEP: A flexible free-software package for electromagnetic simulations by the FDTD method. Computer Physics Communications, 2010

35 A Shinya, S Mitsugi, E Kuramochi, M Notomi, Ultrasmall multi-channel resonant-tunneling filter using mode gap of width-tuned photonic-crystal waveguide. Optics Express, 2005

36 B S Song, T Nagashima, T Asano, S Noda, Resonant-wavelength tuning of a nanocavity by subnanometer control of a two-dimensional silicon-based photonic crystal slab structure. Applied Optics, 2009

37 Y Z Liu, R J Liu, C Z Zhou, D Z Zhang, Z Y Li, Gamma-Mu waveguides in two-dimensional triangular-lattice photonic crystal slabs. Optics Express, 2008

38 C Z Zhou, Y Z Liu, Z Y Li, Waveguide bend of 90° in two-dimensional triangular lattice silicon photonic crystal slabs. Chinese Physics Letters, 2010

39 H H Tao, C Ren, Y Z Liu, Q K Wang, D Z Zhang, Z Y Li, Near-field observation of anomalous optical propagation in photonic crystal coupled-cavity waveguides. Optics Express, 2010

40 K Hennessy, A Badolato, M Winger, D Gerace, M Atatüre, S Gulde, S Fält, E L Hu, A Imamoglu, Quantum nature of a strongly coupled single quantum dot-cavity system. Nature, 2007

41 Y Akahane, T Asano, B S Song, S Noda, High-Q photonic nanocavity in a two-dimensional photonic crystal. Nature, 2003

42 C Z Zhou, C Wang, Z Y Li, Fabrication and spectra-measurement of high Q photonic crystal cavity on silicon slab", Acta Physica Sinica, 2012

43 C Ren, J Tian, S Feng, H H Tao, Y Z Liu, K Ren, Z Y Li, B Y Cheng, D Z Zhang, H F Yang, High resolution three-port filter in two dimensional photonic crystal slabs. Optics Express, 2006

44 Y Z Liu, S A Feng, J Tian, C Ren, H H Tao, Z Y Li, B Y Cheng, D Z Zhang, Q Luo, Multichannel filters with shape designing in two-dimensional photonic crystal slabs. Journal of Applied Physics, 2007

45 Y Z Liu, R J Liu, S A Feng, C Ren, H F Yang, D Z Zhang, Z Y Li, Multichannel filters via Γ-M and Γ-K waveguide coupling in two-dimensional triangular-lattice photonic crystal slabs. Applied Physics Letters, 2008

46 L Gan, Y Z Liu, J Y Li, Z B Zhang, D Z Zhang, Z Y Li, Ray trace visualization of negative refraction of light in two-dimensional air-bridged silicon photonic crystal slabs at 1.55 μm. Optics Express, 2009

47 H M Nguyen, M A Dundar, R W Van Der Heijden, E Van Der Drift, W J M Salemink, H W M Rogge, S Caro, J. Compact Mach-Zehnder interferometer based on self-collimation of light in a silicon photonic crystal. Optics Express, 2010

48 T P White, C M De Sterke, R C Mcphedran, L C Botten, Highly efficient wide-angle transmission into uniform rod-type photonic crystals. Applied Physics Letters, 2005

49 L Gan, F Qin, Z Y Li, Broadband large-angle self-collimation in two-dimensional silicon photonic crystal", Optics Letters, 2012

50 A E Serebryannikov, One-way diffraction effects in photonic crystal gratings made of isotropic materials. Physical Review B: Condensed Matter and Materials Physics. 2009

51 C Wang, C Z Zhou, and Z Y Li, On-chip optical diode based on silicon photonic crystal heterojunctions. Optics Express, 2011

52 L Feng, M Ayache, J Huang, Y L Xu, M H Lu, Y F Chen, Y Fainman, and A Scherer, Nonreciprocal light propagation in a silicon photonic circuit. Science, 2011

53 S Fan, R Baets, A Petrov, Z Yu, J D Joannopoulos, W Freude, A Melloni, M Popovic, M Vanwolleghem, D Jalas, M Eich, M Krause, H Renner, E Brinkmeyer, and C R Doerr, Comment on "Nonreciprocal light propagation in a silicon photonic circuit". Science, 2012b.

54 L Feng, M Ayache, J Huang, Y L Xu, M H Lu, Y F Chen, Y Fainman, and A Scherer, Response to comment on "Nonreciprocal light propagation in a silicon photonic circuit". Science, 2012c.

55 C Wang, X L Zhong, and Z Y Li, Linear and passive silicon optical isolator. Scientific Reports, 2012

56 B S Song, T Asano, Y Akahane, Y Tanaka, and S Noda, Transmission and reflection characteristics of in-plane hetero-photonic crystals. Applied Physics Letters 2004

57 C Wang, C Z Zhou, and Z Y Li, Creation of stop band by introducing parallel-hetero perturbation in two-dimensional photonic crystal waveguides Journal of Optics, 2011

58 C Wang, and Z Y Li, Cavities without confinement barrier in incommensurate photonic crystal superlattices EPL, 2012

CHAPTER 9

Photonic Nanostructures Design and Optimization for Solar Cell Application

Qian Liu [1], Eric Sandgren [1], Miles Barnhart [2], Rui Zhu [2] and Guoliang Huang [2,*]

[1] Department of System Engineering, University of Arkansas at Little Rock, 2801 S University Ave, Little Rock, AR 72204, USA

[2] Department of Mechanical & Aerospace Engineering, University of Missouri, E3422, Lafferre Hall, Columbia, MO 65211, USA

ABSTRACT

In this paper, a semiconducting photonic nanostructure capable of wide range absorption and tunable optical resonance has been designed with a proposed theoretical optimization model. The design consists of ZnO/CdS core-shell nanowire arrays as well as multilayer thin films that act to absorb incident electromagnetic (EM) waves over a broad frequency range. Theoretical, as well as numerical, studies of the nanostructure inside a solar cell plate have been conducted in order to validate the proposed microstructural design. Excellent energy absorption rates of EM waves have been achieved in the high frequency range by using the optical resonance of the nanowire array. By combining multilayer thin film with the core-shell nanowire in the unit cell of a photonic solar cell, a broadband high absorption has been achieved. Moreover, the geometry of the proposed photonic nanostructure is obtained through the implementation of a genetic algorithm. This avoids local minima and an optimized absorption rate of ~90% over the frequency range of 300 to 750 THz has been obtained in the solar cell.

KEYWORDS

solar cell; photonic nanostructures; optical resonance; genetic algorithm; structural optimization

1. INTRODUCTION

Light trapping via the chemical and geometric design of solar cells promises to dramatically increase solar cell efficiency by increasing light absorption without increasing the exciton or free carrier collection length. Various approaches have been proposed to enhance the overall solar cell efficiency in the thin film or nano-based solar cells. These include: putting a photonic crystal or distributed Bragg reflector in the back of thin film solar cells [1,2,3], using a disperse carbon interlayer and textured photonic crystal backside reflector to prompt back reflected effects [4,5], and combining holey double-layer photonic crystals with ultrathin c-Si solar cells [6]. Recently, solar cells utilizing nanowire arrays, overlapping nanospheres, nano-colloids, and self-assembled silver nanoparticles have shown promise by exhibiting enhanced light trapping effect and greater energy absorption rates than conventional planar semiconductors [7,8,9,10]. Grandidier *et al.* [8] experimentally and theoretically demonstrated efficiency enhancements in solar cells from dielectric nanosphere arrays. Mendes *et al.* [9] developed novel plasmonic back-reflector structures using spherical gold colloids for pronounced far-field scattering with a remarkable broadband enhancement. Morawiec *et al.* [10] demonstrated the broadband absorption enhancement of thin solar cells due to the light scattering effects of plasmonic back reflectors composed of self-assembled silver nanoparticles. Specifically, nanowire arrays have drawn a great deal of attentions due to their optical absorption properties which can enhance the light trapping effect. Hu and Chen [11] demonstrated that Si nanowire is capable of significantly improving light absorption when compared to a typical Si thin film within a high frequency region. ZnO has also been studied extensively as a semiconductor in solar cell applications due to its wide band gap and high electron mobility [7,12,13]. Ordered arrays of nanowires can benefit from diffraction effects, collective resonances, and changes in the optical density of states which are absent in single nanowire or random arrays of nanowires. Garnett and Yang [14] measured the photocurrent of solar cells made with ordered silicon nanowire arrays with different silicon absorber thicknesses to determine the light trapping path length enhancement factor. The current study of nanowire arrays has been focused mainly on Si and ZnO type compositions. Myung *et al.* [15] synthesized a high density ZnO/CdS core-shell nanowire array by depositing the CdS outer layers on pre-grown, vertically-aligned ZnO nanowire, through chemical vapor deposition. Seol *et al.* [16] created a CdSe/CdS/ZnO nanowire array through a three-step solution-based method which was used as a photoanode of a quantum dot sensitized solar cell capable of generating a maximum power conversion efficiency of 4.15%. Although experimental observations of the enhanced absorptions have been reported in those solar cells with nanowire arrays, few theoretical and simulation works have been conducted which can be very useful to provide guidelines for designing high efficiency solar cell plates with outstanding light trapping effect.

Moreover, photonic nanowire arrays assembled with thin film structures have exhibited an increase in light trapping in a large frequency range. Li *et al.* [17] found that optical absorption can be greatly enhanced with the addition of a Si nano-pillar to a Si thin film owing to suppressed reflection and transmission. Lin *et al.* [18] demonstrated an improvement in the photovoltaic performance of a ZnO/P_3HT hybrid nanostructure through interface molecular modification on the ZnO nanowire surface. Wu *et al.* [19] reported a 74% increase in efficiency of ZnO nanowire dye-sensitized (DSSC) thin film solar cells. Mokkapati *et al.* [20] provided criteria for optimizing light trapping of metal nanoparticles in periodic arrays in a long wavelength range, allowing for a higher order diffraction mode while maintaining the highest possible fill factor. Therefore, designing a photonic crystal unit cell that combines multilayer thin film structures and multi-shell nanowires with optimized geometric properties can lead to solar cells with a broadband, high-energy absorption rate.

The objective for this paper is to investigate a solar cell consisting of multilayer thin film structures, as well as periodic core-shell nanowire photonic crystals, as an optical absorber. First, theoretical and numerical analyses are carried out on a simple five-layered thin film solar cell. Next, a microstructural design consisting of ZnO/CdS nanowire arrays and multilayer thin films is proposed. Light absorbing performance of the proposed design is evaluated numerically. In order to yield the highest absorption rate possible in both low and high frequency ranges, non-linear optimization is carried out to generate structural geometries that exhibit excellent absorption performance in a broad frequency range.

2. THEORETICAL ANALYSIS OF ONE-DIMENSIONAL SOLAR CELL WITH MULTILAYER NANO FILMS

To achieve broadband optical absorption in real structures such as thin solar cell plates, photonic nanostructures designed with complex geometries are considered. Theoretical investigation of multilayer nanoscale thin film inside the solar cell plate can be a good starting point to understand the energy absorption inside the solar cell nanostructures. In addition, the periodicity of the solar cell nanostructures can be designed along with the reflection and transmission rates. Furthermore, numerical simulations should be performed to yield the optimal structural parameters and, therefore, increase the photon absorption rate.

First, theoretical investigations of the one-dimensional (1D) multilayer solar cell plate are conducted, with the goal of achieving a guideline for the microstructural design that is focused on absorption in a high frequency region. This model will then be extended to the three-dimensional (3D) case in the next section to investigate and design a solar cell with both nanowire arrays and composite thin films. In order to understand the underlying physical mechanism for photon absorption in optical resonance, analytical modeling as well as numerical simulations are carried out. An analytical model of the multilayer planar solar cell is initially constructed by using the matrix transfer formulation to calculate the reflection and transmission in a 1D multilayered planar structure, as shown in Figure 1. The structure consists of $(N-2)$ layers sandwiched together by two semi-infinite layers (layer 1 and layer N).

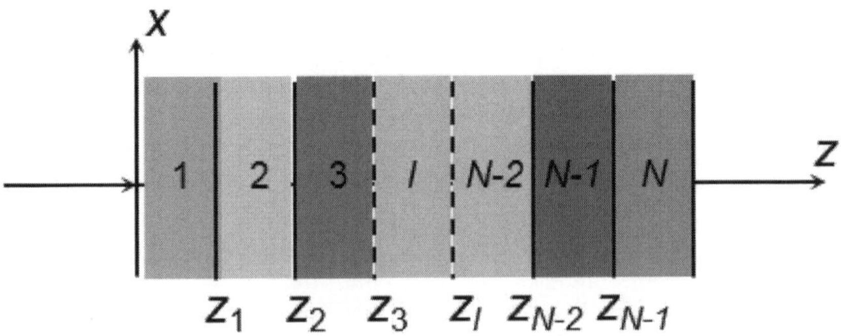

Figure 1. 1D multilayer planar structure.

The relative permittivity, permeability, and refractive index of the l-th layer is denoted by ε_l, μ_l and n_l ($l = 1, 2,\ldots, N$), respectively. Each layer of the multilayer structure is a medium with complex permittivity and magnetic permeability given by [21]:

$$\varepsilon_l^* = \varepsilon_l + i\varepsilon_l' \qquad \mu_l^* = \mu_l + i\mu_l^i \tag{1}$$

where the imaginary portion represents optical absorption behavior in the medium and the dielectric loss tangent can be written as:

$$tan\delta_l = \frac{\varepsilon_l'}{\varepsilon_l} \tag{2}$$

In this paper, we use relative permittivity when carrying out analysis and simulations, which is defined as the ratio of dielectric permittivity over the permittivity of vacuum. For the transverse electric (TE) wave, the electric and magnetic fields in the region l are given by [22]:

$$\boldsymbol{E}_l = \hat{\boldsymbol{y}} A_l e^{ik_x x + ik_{lz}(z - z_{l-1}) - i\omega t} + \hat{\boldsymbol{y}} B_l e^{ik_x x + ik_{lz}(z - z_{l-1}) - i\omega t}$$
$$\boldsymbol{H}_l = \left[-\frac{k_{lz}}{\omega\mu_0\mu_l}\hat{\boldsymbol{x}} + \frac{k_x}{\omega\mu_0\mu_l}\hat{\boldsymbol{z}}\right] A_l e^{ik_x x + ik_{lz}(z - z_{l-1}) - i\omega t} + \left[\frac{k_{lz}}{\omega\mu_0\mu_l}\hat{\boldsymbol{x}} + \frac{k_x}{\omega\mu_0\mu_l}\hat{\boldsymbol{z}}\right] B_l e^{ik_x x - ik_{lz}(z - z_{l-1}) - i\omega t} \tag{3}$$

where $z_0 = z_l = 0$, $z_l = z_{l-1} + d_l$ and d_l is the thickness of the l-th layer ($l = 2, \ldots, N-1$). The constants A_l and B_l represent wave components with the propagating velocity components along the positive and negative z directions, respectively. The variables ε_0 and μ_l are the permittivity and permeability of vacuum, respectively. The total fields for transverse magnetic (TM) waves can then be obtained by interchanging El→Hl, Hl→−El, and μl↔εl in Equation (3). The constants $A_l$and B_l can be determined by using interfacial conditions at z=zl (l = 1, 2,…, N − 1), which lead to:

$$\begin{bmatrix} A_l \\ B_l \end{bmatrix} = \boldsymbol{P}_l \boldsymbol{D}_l^{-1} \boldsymbol{D}_{l+1} \begin{bmatrix} A_{l+1} \\ B_{l+1} \end{bmatrix} \tag{4}$$

where D−1l is the inverse matrix of Dl. The matrices Pl and Dl are given by

$$\boldsymbol{P}_l = \begin{bmatrix} e^{\phi_l} & 0 \\ 0 & e^{-\phi_l} \end{bmatrix} \tag{5}$$

$$\boldsymbol{D}_l = \begin{bmatrix} 1 & 1 \\ \eta & -\eta \end{bmatrix} \tag{6}$$

respectively. In Equations (5) and (6),

$$\phi_1 = 0 \qquad \phi_l = \begin{cases} (\omega/c)d_l\alpha_l, & k_x^2 > k_l^2 \\ -i(\omega/c)d_l\alpha_l, k_l^2 > k_x^2 \end{cases} (l = 2,3, \dots, N-1) \tag{7}$$

And

$$\eta = \begin{cases} i\gamma_l, k_x^2 > k_l^2 \\ \gamma_l, k_l^2 > k_x^2 \end{cases} \tag{8}$$

with γl=αl/μl for a TE wave and γl=αl/εl for a TM wave. Note that

$$\alpha_l = \begin{cases} \sqrt{(n_1 sin\theta_1)^2 - n_l^2}, k_x^2 > k_l^2 \\ \sqrt{n_l^2 - (n_1 sin\theta_1)^2}, k_l^2 > k_x^2 \end{cases} \tag{9}$$

where $n_l^2 = \varepsilon_l$ ($l = 1, 2, \ldots, N - 1$) is the refraction index. From Equation (4), we can derive the following relation

$$\begin{bmatrix} A_1 \\ B_1 \end{bmatrix} = \boldsymbol{M} \begin{bmatrix} A_N \\ B_N \end{bmatrix} \tag{10}$$

where the 2×2 matrix $\boldsymbol{M} = \prod_{l=1}^{N-1} \boldsymbol{P}_l \boldsymbol{D}_l^{-1} \boldsymbol{D}_{l+1}$.

One can set $B_N = 0$, since there is no reflection in the region N. Therefore, the transmission and reflection coefficients can be calculated as $\tilde{t} = 1/M_{11}$ and $\tilde{r} = M_{21}/M_{11}$, respectively. M_{ij} $(i, j = 1, 2)$ are the components of the 2 × 2 matrix **M**. Finally, the power transmittance T and reflectance R can be calculated as:

$$T = \frac{n_N cos\theta_N}{n_1 cos\theta_1} \widetilde{tt^*} \tag{11}$$

$$R = \widetilde{rr^*} \tag{12}$$

respectively, where $\theta_N = \sin^{-1}(n_1 \sin\theta_1 / n_N)$.

In Figure 2, a five-layer thin film solar cell with vacuums as the top and bottom outer layers are studied with different angles of incidence. Both a theoretical model and finite element (FE) model are used to calculate the reflection, transmission, and absorption rates of the five-layer thin film. The relative permittivity and dielectric loss tangents of ZnO, CdS, and CZTS layers are all frequency-dependent and determined through experimental measurement. Figure 3a,b demonstrate the dielectric loss tangents and the real parts of the relative permittivity of the three layers, respectively.

Figure 2. Five-layer thin film solar cell under different angles of incidence.

The thicknesses of ZnO, CdS, and CZTS layers are 400, 200 and 200 nm, respectively. Three different angles of incidence (0°, 30°, and 60°) are studied. In order to validate the analytical model, a numerical simulation is also conducted on the five-layer thin film with the same dispersive material parameters by using Ansoft HFSS14. In Figure 4, the reflection, transmission, and absorption results obtained from the theoretical model and the FE model are plotted as solid curves and small triangles, respectively. Excellent agreement between the two results can be observed, which confirms that the developed theoretical model can describe the exact wave behavior in the multilayer thin film solar cell.

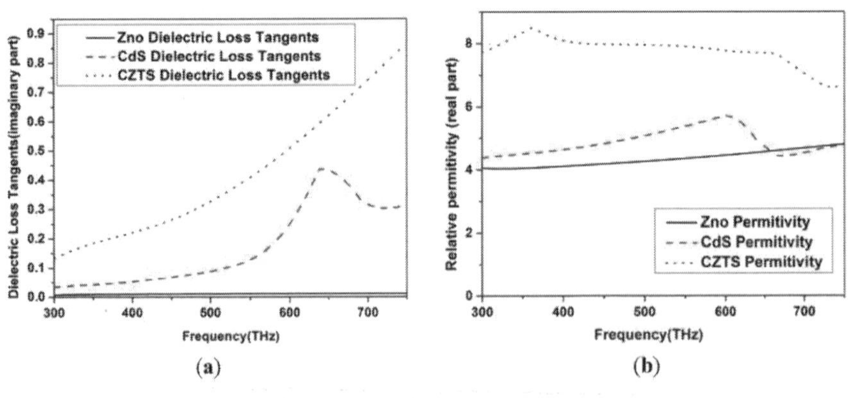

Figure 3. (**a**) Imaginary parts of the dielectric loss tangent. (**b**) Real parts of the relative permittivity of ZnO, CdS, and CZTS layers.

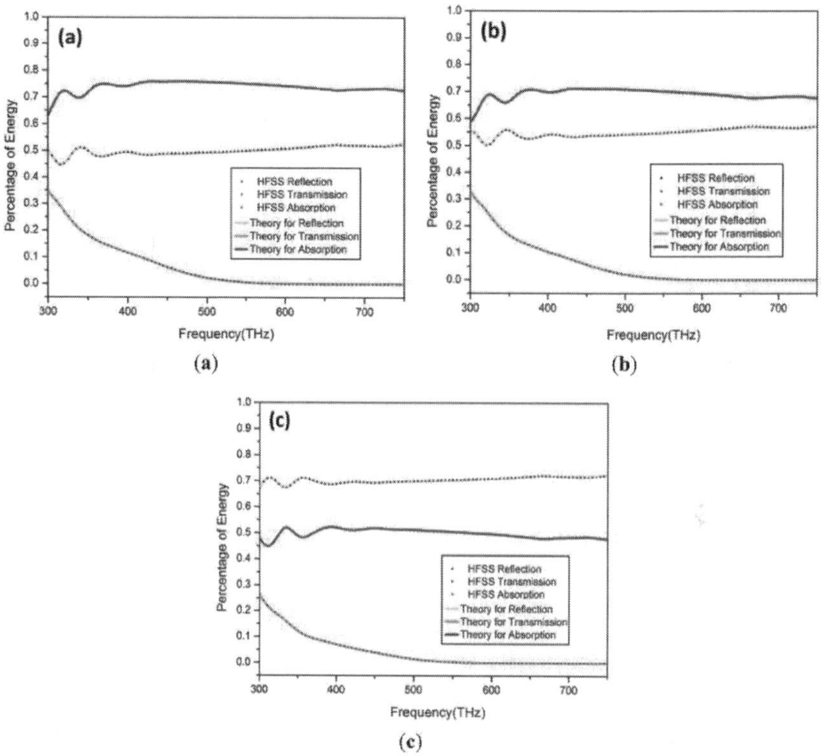

Figure 4. Theoretical and numerical results of the reflection, transmission and absorption of the five-layer thin film solar cell under (**a**) 0°. (**b**) 30°. (**c**) 60° angle of incident.

Also, by comparing the results of the three different angles of incidence, it can be found that the reflection increases with larger angle of incident while the few changes in transmission can be observed especially in the high frequency range. The absorptions for 0° and 30° incidences are almost the same, while decreased absorption can be found when the angle of incidence increases to 60°.

3. NUMERICAL SIMULATION AND OPTIMIZATION ON SOLAR CELL CONSISTING OF CORE-SHELL NANOWIRE ARRAYS AND MULTI-LAYER THIN FILMS

Nanowire arrays have demonstrated unusual optical absorption properties in solar cell applications. Current studies of nanowires are mostly concerned with Si and ZnO compositions. Hu and Chen's study [11] mentioned that nanowire arrays with moderate filling ratio could exhibit greater absorption rates and lower reflectance compared with thin films, especially in a high frequency regime. The nanostructure design is of great concern in the research and development of novel solar cell structures. For nanostructures with complex geometry, the aids of numerical methods such as the finite element method (FEM) are required. In our study, a core-shell nanowire is used in order to obtain enhanced absorption. The inner core is ZnO and the outer shell is made of CdS. The core-shell nanowire is then situated within a vacuum. Simulations are conducted by using Ansoft High Frequency Structure Simulator (HFSS) with a Floquet port feature, which is capable of modeling the oblique incidence and polarization of the wave at the boundary of the periodic-planar structure. A photonic structure is created by using MATLAB in combination with HFSS's Visual Basic scripting. Results of the magnitude and phase of the reflected and transmitted waves, as well as absorption rate through the periodic nanostructures, are then obtained. Figure 5 shows the simulation result of the energy flux density of a unit cell with core-shell nanowire structure and Figure 6 shows the absorption rate of the core-shell nanowire structure; the coefficient describes intensity or total power percentage absorbed relative to an incident wave. In Figure 6, we also draw the reflection and transmission coefficient curves; these two amplitude coefficients are represented by lower case r and t in Equations (11) and (12).

It can be observed from Figure 6 that obviously higher absorption and relatively lower reflectance can be achieved at a high frequency range compared with those at a low frequency range. The influence of the optical resonance on the increased absorption can be found from 600 to 750 THz, as shown in Figure 6. However, it will be ideal if the absorption rate is consistently high across a broad frequency range. To accomplish this, the original core-shell model needs to be modified through the optimization of structural dimensions, as well as the material compositions inside each unit cell of the nanowire array.

Furthermore, a nanowire array combined with a nanofilm can significantly enhance the optical absorption of a solar cell in a broad frequency range and has become a promising direction for future research of high-efficiency solar cells. By adding the Si nanowires on Si thin film, Li *et al.* [12] found that optical absorption of the combined structure could be greatly enhanced. To begin the microstructural design of the optimized 3D solar cell, we first consider a periodic array of double shelled (ZnO/CdS/CZTS) nanowire structures that are embedded in a multilayer thin film. Details of the initial unit cell design can be found in Figure 7. A ZnO nanowire is first grown on the top of a ZnO thin film, which is then encapsulated by a layer of CdS via vapor deposition. Finally, a top layer of CZTS is added via vapor deposition to envelope the middle layer of CdS in the unit cell. Analysis on the initial design will be conducted in order to get baseline absorption rates prior to optimization.

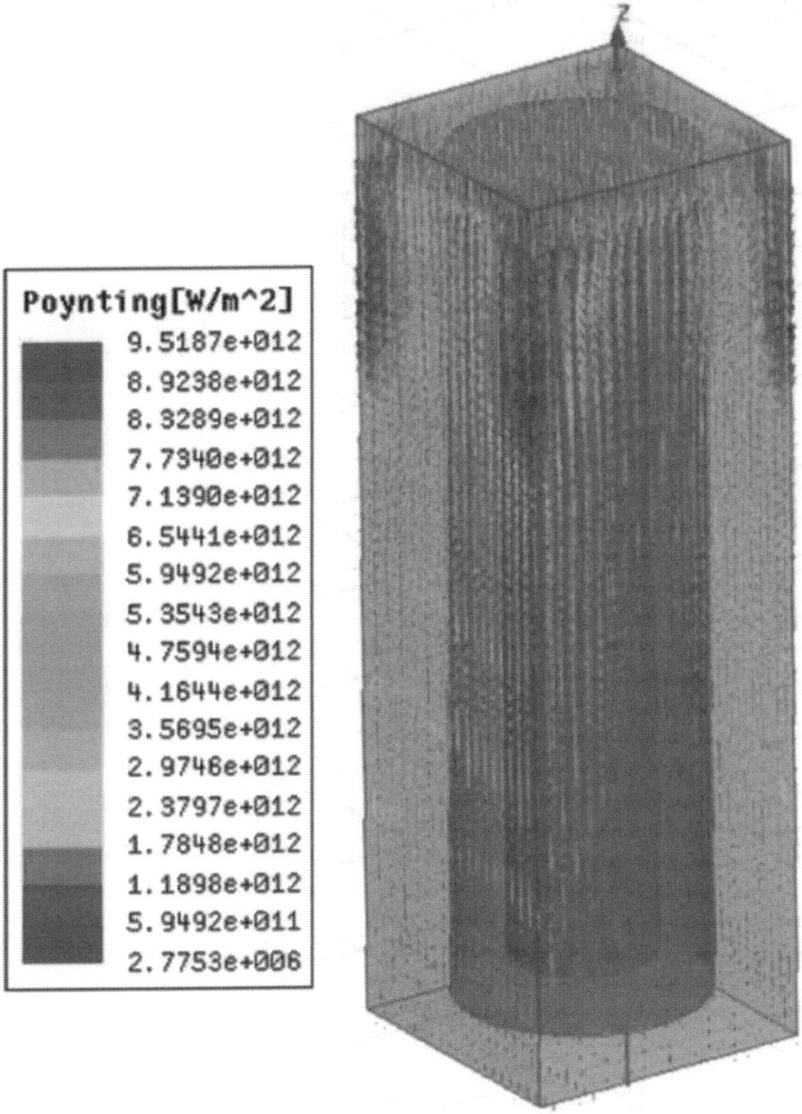

Figure 5. The energy field of the unit cell with core-shell nanowire structure.

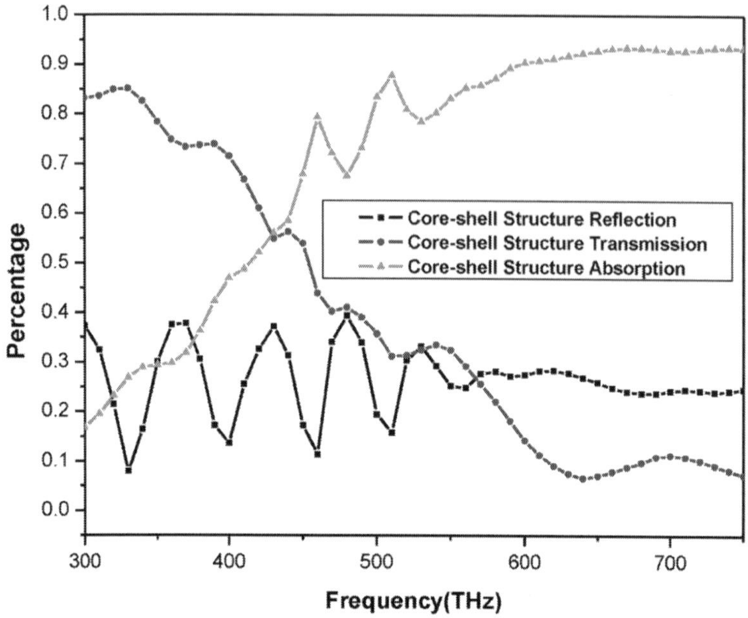

Figure 6. Absorption rate of the core-shell nanowire structure.

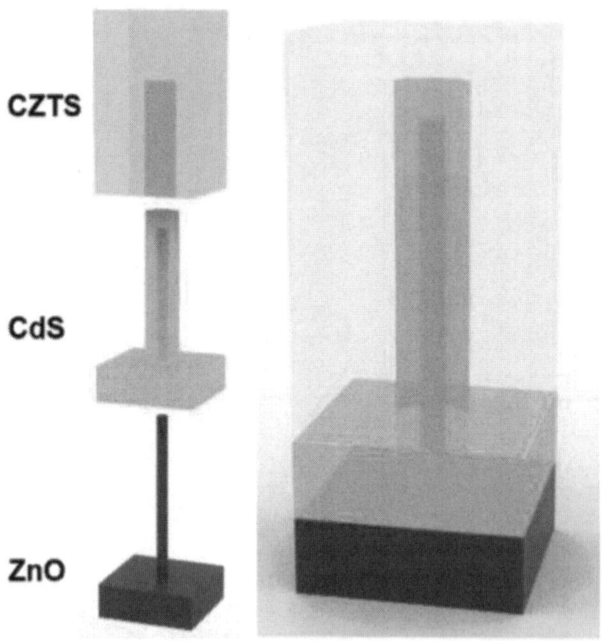

Figure 7. Model of the proposed nanowire-nanofilm unit cell configuration.

The 3D solar cell is created and the optical wave simulation in the nanowire and layered medium is conducted by using Ansoft HFSS14. Periodicity of the structure is utilized in the simulation. As a result, analysis is only carried out on a single unit cell where periodic boundary conditions are applied. The Floquet port is used exclusively with planar-periodic structures in HFSS. Chief examples are planar phased arrays and frequency-selective surfaces which may be idealized as infinitely large. The analysis of the infinite structure is then accomplished by analyzing a unit cell. Linked boundaries most often form the side walls of a unit cell but, in addition, a boundary condition is required to account for the infinite space above. The Floquet port is designed for this purpose. The Floquet port is closely related to a wave port in that a set of modes ("Floquet modes") represent the fields on the port boundary. Fundamentally, Floquet modes are plane waves with the propagation direction set by the frequency, phasing, and geometry of the periodic structure. Just like wave modes, Floquet modes also have propagation constants and experience cut-off at low frequency. Ansoft HFSS is used, of which the Floquet port can also deal with the oblique incident wave with certain polarization illuminating the periodic material structure problem. This was done in order to get the magnitude and phase of the reflection and transmission waves. The HFSS solution includes a model decomposition of which all information is obtained in the form of an S-matrix interrelating the Floquet modes. Figure 8a shows the dimensions of a single unit cell in the simulation. The three layers in the middle of the FE model are the dispersive materials CZTS, CdS, and ZnO from top to bottom, respectively. The value of permeability are all equal to unity ($\mu = 1$) since the materials are nonmagnetic over the entire optical frequency range of interest. The permittivity of vacuum is 1 and the others' permittivity and dielectric loss tangents are all dispersive, experimentally-determined optical constants of those material components and are used in the FE simulation. The thicknesses of the three thin films are set to 400, 200, and 200 nm for CZTS, CdS, and ZnO, respectively. The incident light is set normal to the film substrate, *i.e.*, it is parallel to the nanowires. Periodic conditions are employed at a pair of surfaces parallel to the x-z and y-$z$planes by using the master/slave boundary. The final structure is realized through infinitely extending the structural units having the height of 1800 nm and a width dimension equal to the nanowire array periodicity, as indicated by the d_l inFigure 8a,b showing the meshed unit cell which is sandwiched between two bodies of vacuum with finite length.

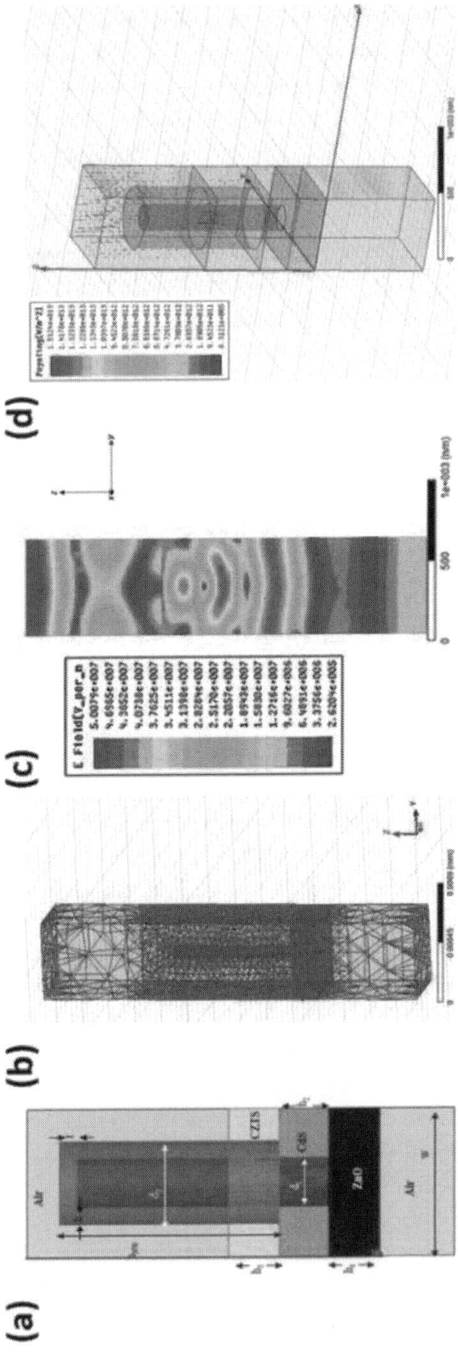

Figure 8. (**a**) Dimensions of a single unit cell. (**b**)The meshed FE model. (**c**) The simulated electric field flux. (**d**) The simulated Poynting energy field in the unit cell.

The optical information is obtained from the spatial distribution of the electric field flux which results from the interaction between the structure and the incident light, as shown in Figure 8c. The frequency range of the incident light varies between 300 to 750 THz corresponding to the wavelength from 400 to 1000 nm. These values are selected because they occupy the frequency spectrum of solar radiation which is of particular interest. The results of Poynting flux distribution is shown inFigure 8d. It can be found from Figure 8c,d that the electric field in the core-shell nanowire regime is enhanced and energy is gathered by the nanostructure due to light trapping effect. Therefore, very little energy is transmitted out, but still a large of part is reflected back by the nanostructure.

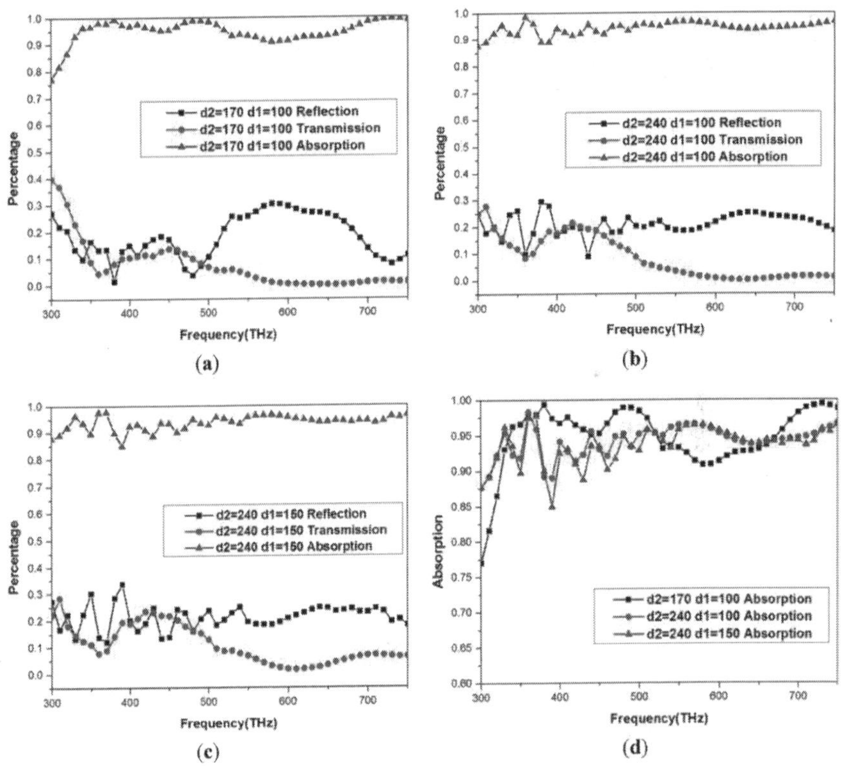

Figure 9. FEM results for absorption rates of varies nanowire diameters.

In order to optimize the proposed unit cell, a genetic algorithm (GA) is used because its interactive process can avoid local minima and, therefore, is an ideal optimization method for this particular problem. An optimal photonic structure is created using MATLAB, in combination with HFSSs Visual Basic scripting, where optimization is carried out via a

GA on the proposed design. The GA mimics naïve evolution where the objective function, or in this case maximum energy absorption, is used as a measure of fitness for the system. The GA is a three-step optimization process that implements selection, crossover, and mutation over generations (iterations) in order to seek out an optimal structure topology. Through the iterative process geometric parameters, including the unit cell width, w, thin film thicknesses h_1, h_2, h_3, and the height, h, as well as the diameters, d_1, d_2, and the unit cell width, w, of the core/shell are randomly selected until a global minimum is found (optimal fitness). In this case, the GA is utilized to find the geometric parameters that result in the maximum energy absorption over the entire frequency spectrum. This is accomplished by minimizing the reciprocal of the maximum energy absorption. After structural optimization is carried out, excellent absorptions with an average rate of approximately 90% can be observed over a broad high frequency range from 300 to 750 THz, as shown in Figure 9. It can be found that optimization results are achieved in the optical spectrum, 400–1000 nm (300–750 THz), where most energy comes from and different optimized designs have different maximum absorption frequency ranges. For example, the optimized structure with $d_2 = 170$ $d_i = 100$ can reach the maximum absorption rate in two frequency ranges of 350–500 THz and 650–750 THz, while in 500–650 THz the absorption rate of this design is the least among all three optimized designs. The obtained optimized structures can be used in solar cells with specifically-emphasized operating frequency ranges.

4. CONCLUSIONS

In summary, this paper has demonstrated a broadband, high energy absorption rate of the proposed photonic crystal solar cell that utilizes optical resonance-enhanced light trapping. First, analytic and numerical analysis are carried out on a simple five-layered thin film solar cell in which very good agreement between both models can be found. Next, a photonic crystal design consisting of ZnO/CdS core-shell nanowire arrays is proposed. The energy absorption rate of the proposed photonic nanostructure has been evaluated. An excellent absorption rate is obtained in the high frequency range but ineffective performance is found in the low frequency region. Finally, non-linear optimization is carried out via a genetic algorithm on a photonic crystal solar cell consisting of multilayer thin film structures and periodic core-shell nanowire arrays. The optimized structure has demonstrated excellent performance over a broad frequency range (300 to 750 THz) with an average absorption ratio of 90%. The theoretical and numerical studies on the light trapping and energy absorption in a photonic crystal with multilayer thin film structures and multi-shell nanowire arrays can be very useful to provide guidelines for designing high efficiency solar cell plates. Furthermore, the optimized design of the photonic crystal unit cell can be quite useful in designing solar cells with broadband high energy absorption rates.

ACKNOWLEDGMENTS

The authors would like to thank Dr. Xiaoming Zhou from Beijing Institute of Technology for his comments and discussions on the theoretical model. This research was supported by the National Science Foundation under award No. EPS-1003970.

AUTHOR CONTRIBUTIONS

Q.L. has performed the theoretical and numerical investigations. M.B., R.Z., E.S. and G.L. have helped in the discussions. G.L. has supervised the research. All authors have contributed to the final version of the manuscript.

CONFLICTS OF INTEREST

The authors declare no conflict of interest.

REFERENCES

1 Bermel, P.; Luo, C.; Zeng, L.; Kimerling, L.C.; Joannopoulos, J.D. Improving thin-film crystalline silicon solar cell efficiencies with photonic crystals. *Opt. Express* **2007**, *15*, 16986–17000.

2 Zhou, D.; Biswas, R. Photonic crystal enhanced light-trapping in thin film solar cells. *J. Appl. Phys.* **2008**, *103*.

3 Chutinan, A.; Li, C.W.W.; Kherani, N.P.; Zukotynski, S. Wave-optical studies of light trapping in submicrometre-textured ultra-thin crystalline silicon solar cells. *J. Phys. D Appl. Phys.* **2011**, *44*.

4 Araújo, A.; Barros, R.; Mateus, T.; Gaspar, D.; Neves, N.; Vicente, A.; Martins, R. Role of a disperse carbon interlayer on the performances of tandem a-Si solar cells. *Sci. Technol. Adv. Mater.* **2013**, *14*.

5 Zeng, L.; Bermel, P.; Yi, Y.; Alamariu, B.A.; Broderick, K.A.; Liu, J.; Kimerling, L.C. Demonstration of enhanced absorption in thin film Si solar cells with textured photonic crystal back reflector. *Appl. Phys. Lett.* **2008**, *93*.

6 Mallick, S.B.; Agrawal, M.; Peumans, P. Optimal light trapping in ultra-thin photonic crystal crystalline silicon solar cells. *Opt. Express* **2010**, *18*, 5691–5706.

7 Yi, G.C.; Wang, C.; Park, W.I. ZnO nanorods: Synthesis, characterization and applications. *Semicond. Sci. Technol.* **2005**,*20*.

8 Grandidier, J.; Weitekamp, R.A.; Deceglie, M.G.; Callahan, D.M.; Battaglia, C.; Bukowsky, C.R.; Atwater, H.A. Solar cell efficiency enhancement via light trapping in printable resonant dielectric nanosphere arrays. *Phys. Status Solidi***2013**, *210*, 255–260.

9 Mendes, M.J.; Morawiec, S.; Mateus, T.; Lyubchyk, A.; Águas, H.; Ferreira, I.; Crupi, I. Broadband light trapping in thin film solar cells with self-organized plasmonic nano-colloids. *Nanotechnology* **2015**, *26*.

10 Morawiec, S.; Mendes, M.J.; Filonovich, S.A.; Mateus, T.; Mirabella, S.; Águas, H.; Crupi, I. Broadband photocurrent enhancement in a-Si: H solar cells with plasmonic back reflectors. *Opt. Express* **2014**, *22*, A1059–A1070.

11 Hu, L.; Chen, G. Analysis of optical absorption in silicon nanowire arrays for photovoltaic applications. *Nano Lett.***2007**, *7*, 3249–3252.

12 Peng, Q.; Qin, Y. ZnO nanowires and their application for solar cells. In *Nanowires—Implementations and Applications*; Hashim, A., Ed.; INTECH Open Access Publisher: Rijeka, Croatia, 2011; pp. 158–178.

13 Xia, Y.; Yang, P.; Sun, Y.; Wu, Y.; Mayers, B.; Gates, B.; Yin, Y.; Kim, F.; Yan, H. One-dimensional nanostructures: synthesis, characterization, and applications. *Adv. Mater.* **2003**, *15*, 353–389.

14 Garnett, E.; Yang, P. Light trapping in silicon nanowire solar cells. *Nano Lett.* **2010**, *10*, 1082–1087.

15 Myung, Y.; Jang, D.M.; Sung, T.K.; Sohn, Y.J.; Jung, G.B.; Cho, Y.J.; Kim, H.S.; Park, J. Composition-Tuned ZnO-CdSSe Core-Shell Nanowire Arrays. *ACS Nano* **2010**, *4*, 3789–3800.

16 Seol, M.; Kim, H.; Kim, W.; Yong, K. Highly efficient photoelectrochemical hydrogen generation using a ZnO nanowire array and a CdSe/CdS co-sensitizer. *Electrochem. Commun.* **2010**, *12*, 1416–1418.

17 Li, J.; Yu, H.; Wong, S.M.; Zhang, G.; Sun, X.; Lo, P.G.Q.; Kwong, D.L. Si nanopillar array optimization on Si thin films for solar energy harvesting. *Appl. Phys. Lett.* **2009**, *95*.

18 Lin, Y.Y.; Lee, Y.Y.; Chang, L.; Wu, J.J.; Chen, C.W. The influence of interface modifier on the performance of nanostructured ZnO/polymer hybrid solar cells. *Appl. Phys. Lett.* **2009**, *94*.

19 Wu, J.J.; Chen, Y.R.; Liao, W.P.; Wu, C.T.; Chen, C.Y. Construction of nanocrystalline film on nanowire array via swelling electrospun polyvinylpyrrolidone-hosted nanofibers for use in dye-sensitized solar cells. *ACS Nano* **2010**, *4*, 5679–5684.

20 Mokkapati, S.; Beck, F.J.; Polman, A.; Catchpole, K.R. Designing periodic arrays of metal nanoparticles for light-trapping applications in solar cells. *Appl. Phys. Lett.* **2009**, *95*.

21 Veselago, V.G. The electrodynamics of substances with simultaneously negative values of ε and μ. *Sov. Phys. Usp.* **1968**, *10*, 509–514.

22 Zhou, X.; Hu, G. Total transmission condition for photon tunnelling in a layered structure with metamaterials. *J. Opt. A Pure Appl. Opt.* **2007**, *9*.

CHAPTER 10

Design and Analysis of Enhanced Modulation Response in Integrated Coupled Cavities DBR Lasers Using Photon-Photon Resonance

Paolo Bardella[1,*], Weng W. Chow[2] and Ivo Montrosset[1]

[1] Dipartimento di Elettronica e Telecomunicazioni, Politecnico di Torino, Corso Duca degli Abruzzi 24, 10129 Torino, Italy

[2] Sandia National Laboratories, Albuquerque, NM 87185, USA

ABSTRACT

In the last few decades, various solutions have been proposed to increase the modulation bandwidth and, consequently, the transmission bit-rate of semiconductor lasers. In this manuscript, we discuss a design procedure for a recently proposed laser cavity realized with the monolithic integration of two distributed Bragg reflector (DBR) lasers allowing one to extend the modulation bandwidth. Such an extension is obtained introducing in the dynamic response a photon-photon resonance (PPR) at a frequency higher than the modulation bandwidth of the corresponding single-section laser. Design guidelines will be proposed, and dynamic small and large signal simulations results, calculated using a finite difference traveling wave (FDTW) numerical simulator, will be discussed to confirm the design results. The effectiveness of the design procedure is verified in a structure with PPR frequency at 35GHz allowing one to obtain an open eye diagram for a non-return-to-zero (NRZ) digital signal up to 80GHz. Furthermore, the investigation of the rich dynamics of this structure shows that with proper bias conditions, it is possible to obtain also a tunable self-pulsating signal in a frequency range related to the PPR design.

KEYWORDS

semiconductor laser diode; monolithically-integrated coupled-cavity DBR lasers; integrated optics; photonic integrated circuits (PICs); direct modulation; small-signal modulation bandwidth; eye diagram; DBR laser; optical injection-locking

1. INTRODUCTION

Semiconductor laser diodes with a wide direct modulation bandwidth represent an important element to fulfill the continuously increasing request for low-cost optical communications systems with a high bit-rate (see, e.g., [1]). Whilst the maximum bit-rate achieved by direct modulated lasers is typically limited by the well-known resonance between carriers and photons (carrier-photon resonance (CPR)) [2], many solutions have been proposed to overcome this restriction; see, for example, [3] for a recent review.

A first mechanism identified to extend the modulation bandwidth is the detuned loading (DL) due to the dispersion effect introduced by a coupled passive cavity [4,5] or by a distributed mirror (DBR [6,7,8] or Distributed Feedback (DFB) [9]) when the lasing mode is properly positioned at a slightly higher wavelength with respect to the minimum threshold gain condition.

A second approach used to extend the lasers dynamic properties is to take advantage, in a properly designed cavity, of the interaction between the lasing mode and an adjacent longitudinal cavity mode. This interaction is made possible by the carrier pulsation introduced by the current modulation applied at the gain section electrode [6,7,10,11,12]. This interaction introduces a resonance in the intensity modulation response at the frequency corresponding to the two cavity modes' separation; such a resonance is frequently called photon-photon resonance (PPR).

Since the PPR usually occurs at a frequency that is much higher than the CPR frequency, the request for an almost flat modulation response implies the need for a proper cavity design to have the PPR at the correct frequency, allowing one to fill the gap between the PPR and CPR peaks of the modulation response. In this condition of modulation bandwidth extension, it is possible to obtain an open eye diagram of an NRZ signal at a greater bit-rate than in the corresponding single-section DBR laser.

An approach that is frequently used to achieve this condition is the introduction of an external feedback to the laser cavity. On this concept, various cavity designs have been studied and realized as, e.g., the complex cavity injection grating (CCIG) [13,14,15,16], DFB with integrated feedback (IFB-DFB) [17] or single-mode cavity with feedback effects [18,19]. Furthermore, the modulation bandwidth extension can be obtained with the injection-locking of a laser to the optical signal of an external source (see, e.g., [20,21,22]). The modulation bandwidth extension has also been obtained exploiting the coupling between two modes in a waveguide supporting two transverse modes [23] or in coupled vertical-cavity surface-emitting lasers (VCSELs) [24,25]. In all of the previously-cited cases, the bandwidth extension by PPR can be used either (1) to improve the dynamic characteristics of a laser that shows a limited modulation bandwidth because of its non-optimal material [26] or cavity [1] properties or (2) to extend the dynamic properties of a device that already exhibits a good modulation response.

The structure investigated in this manuscript consists of two coupled cavity DBR lasers integrated in a single chip, as shown in Figure 1a. This structure can be seen both as an example of a DBR laser with an external feedback from an active cavity or as an integrated injection-locked DBR laser, as has been already presented in [27,28,29,30] depending on the injection level in the two cavities. In this paper, just for simplicity, we will call the master the unmodulated laser cavity and the slave the other one.

The additional contributions in this paper with respect to the previous ones [27,28,29,30] are: the definition of a design procedure allowing one to determine the cavity parameters necessary to achieve a prescribed PPR frequency (fPPR), the validation of the design results by showing the possibility of large signal operation conditions with a clearly open eye diagram at a higher bit-rate with respect to that of the single-cavity configuration and also a more complete mapping of the dynamic characteristic with the demonstration of the existence of conditions of good self-pulsation operation in proper bias conditions.

The paper is organized as follows: First, in Section 2, we will present the cavity design procedure and the obtained PPR frequency maps as functions of the cavity parameters. In Section 3, the results obtained from the simulations of the small and large signal modulation and of self-pulsation operation will be discussed. Finally, in Section 4, we will draw the conclusions.

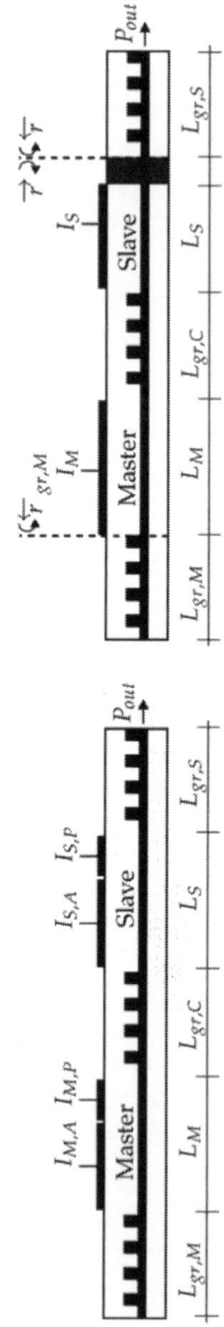

Figure 1. (**a**) Schematics of the coupled-cavity DBR laser. LM and LS are the sum of the lengths of the active and phase-control sections of the two laser cavities. The currents IM,A (IM,P) and IS,A (IS,P) are injected in the active (phase-control) sections of the master (left) and slave (right) laser cavities, respectively. (**b**) Analyzed coupled cavity. The phase tuning sections are replaced by a lumped phase term ϕS. The right dashed line indicates the reference plane at which the equivalent reflectivities $\overrightarrow{r}$ and $\overleftarrow{r}$ are calculated using the transfer matrix method.

2. DESIGN OF THE COUPLED DBR LASER CAVITY

For the composite DBR cavity under consideration (Figure 1a), as well as for the other cases of laser cavity referenced inSection 1, a proper choice of the cavity parameters is the essence for the exploitation of the PPR mechanism between the lasing mode and the nearest neighbor with the requested frequency separation.

The need for a design derives from two competing factors: if the cavity length is reduced to avoid the parasitic effect limiting the laser dynamics, the mode separation is then usually too high to obtain a flat dynamic modulation response using the PPR effect. Therefore, careful design of the cavity is needed in order to control the separation between the two modes that must interact to obtain the modulation extension effect.

The precise mode separation can only be obtained from the above-threshold analysis [16,17], but an analysis at the threshold [12,13] allows a good estimation. Thus, the results found at the threshold may be used to design the structure to be analyzed in the above-threshold regime.

2.1. Master and Slave Lasers' Definition

The first step in the cavity design is the definition of the role of the two lasers. Since we assumed that the slave (right DBR) laser will be modulated, the output power will be therefore extracted from the right mirror. The second assumption is that the master (left DBR) laser has only a small output power on the left side of the cavity in order to maximize its power coupling to the slave laser. Obviously, with these preliminary assumptions, the higher reflectivity at λB will be the peak reflectivity of the left DBR mirror (RL), while the lower one will be the peak reflectivity of the right grating (RR). The peak value of the central mirror reflectivity (RC) will be in between RL and RR; obviously, it will determine the strength of the coupling between the two cavities and, consequently, the frequency splitting fPPR between the resonances of the composite cavity.

2.2. Photon-Photon Resonance Frequency Calculation

We are interested in the case where both cavities are above-threshold with independent current injection in the active regions. First, we assume that the active section of the master (left) cavity is at transparency and that there is gain only in the slave (right) cavity. In this condition, we perform a below-threshold analysis, and we search for the threshold condition for the full coupled cavity structure. To emphasize the role of the two cavities in the lasing mode selection, we assume a wavelength-independent gain function. In this condition, we compute at threshold the gain and the frequency of the lasing mode and of its adjacent ones and the round trip gain (RTG) and phase (RTP) function of the full cavity in the frequency range around the grating Bragg condition. The RTG and RTP functions at the lasing condition are expressed as:

$$\mathrm{RTG}(\lambda) = |\overleftarrow{r}(\lambda) \cdot \overrightarrow{r}(\lambda)|; \quad \mathrm{RTP}(\lambda) = \angle \overleftarrow{r}(\lambda) \cdot \overrightarrow{r}(\lambda)$$

where the equivalent reflectivities $\overleftarrow{r}(\lambda)$ and $\overrightarrow{r}(\lambda)$ are calculated using the transmission matrix approach [2] considering a reference plane placed at the left input of the slave grating (Figure 1b). Details on the calculation of $\overleftarrow{r}(\lambda)$ and $\overrightarrow{r}(\lambda)$ are presented in Appendix A1.

This threshold analysis allows one to obtain the frequency separation between the lasing and the adjacent modes and also gives preliminary information on the mode competition from the gain margin between the lasing mode and the non-lasing ones.

Examples of RTG and RTP functions around the lasing frequency are reported in Figure 2 for three values of the control phase ϕS. In order to qualitatively appreciate the dynamic behavior in the three operation conditions in Figure 2, we present in Figure 3a preview of small signal modulation responses discussed more completely in Section 3 for a suitable choice of the currents injected in the active sections.

When ϕS=0, the lasing mode (red circle) and its closest cavity mode (green square) are separated by 47 GHz (Figure 2a); however, the PPR peak is barely visible in the small signal modulation response (Figure 3a) due to the large gain margin between the lasing mode and the closest one, which is located near the RTG minimum. In this condition, the −3 dB bandwidth only depends on the CPR. When ϕS=π, two modes have RTG =1, and their frequency separation is fPPR=35GHz (Figure 2b): the corresponding small signal modulation response typically shows a strong peak at the PPR frequency (Figure 3b), indicating that the beating between the two modes generates a self-pulsating output power. While this condition could be useful for radio-frequency (RF) photonics applications, it cannot be employed for direct digital modulation.

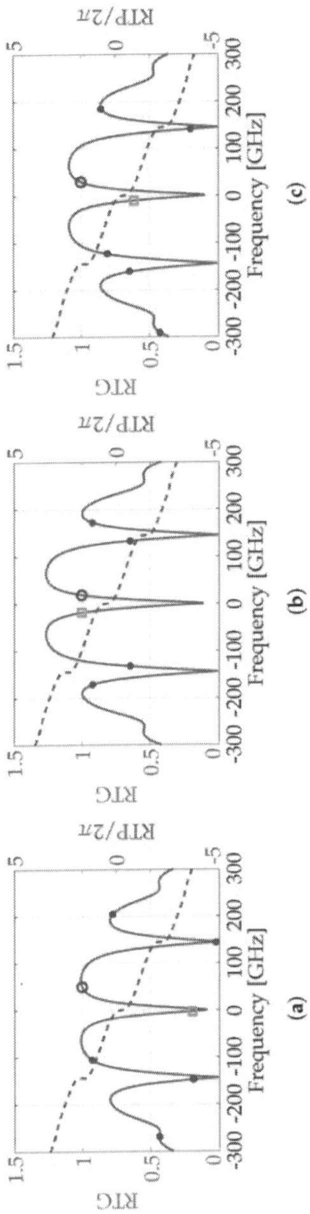

Figure 2. Round trip gain (RTG) (continuous blue line) and phase (RTP) (dashed red line) functions plots at threshold around the Bragg frequency for three different additional phase shifts ϕS in the slave laser cavity. The lasing mode is indicated by the red circle, the closest non-lasing mode by the green square marker and the other non-lasing modes by the blue dots. Lgr,M=181 μm, LM=LS=250 μm, Lgr,C=65 μm, Lgr,S=70 μm, κ=10 0cm−1. **(a)** ϕS=0 and fPPR=47GHz, **(b)** ϕS=π and fPPR=35GHz and **(c)** ϕS=π/2 and fPPR=38GHz.

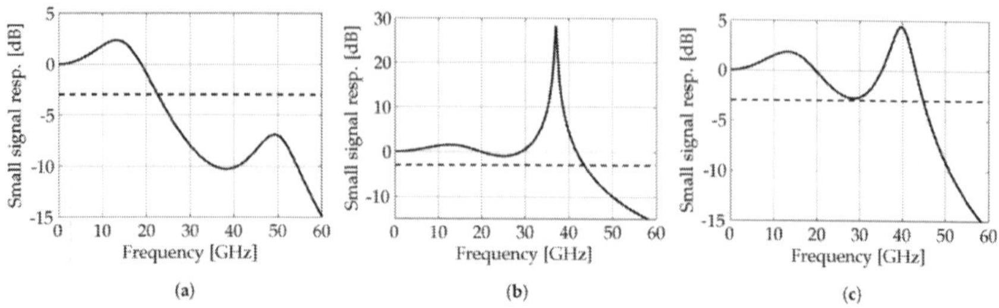

Figure 3. Examples of small signal modulation responses obtained with the finite difference traveling wave (FDTW) model for cavity parameters and phase tuning used in Figure 2. (**a**) $\phi S=0$ and fPPR=49GHz, (**b**) $\phi S=\pi$ and fPPR=37GHz and (**c**) $\phi S=\pi/2$ and fPPR=38GHz.

In order to obtain a modulation response suitable for digital transmission applications, the correct balancing between PPR frequency and the gain margin of the mode closest to the lasing one must be ensured, as in Figure 2c and in the corresponding Figure 3c, where the −3 dB bandwidth is enhanced up to 43GHz. With respect to the case with $\phi S=\pi$, now, the fPPR is increased by 10%, but the PPR peak is strongly reduced. Good large signal operations can be obtained around this phase condition; on the contrary, an excessively large fPPR could generate a gap between the CPR and the PPR peaks, again limiting the −3 dB modulation bandwidth. From our experience, a good choice for the fPPR in order to maximize the bit rate for digital transmission application is fPPR≈3fCPR.

2.3. Coupled Cavities' Design

Different from previous studies [27,28], in our analysis, the distributed characteristics of the DBR mirrors have been considered; therefore, the cavities' parameters for the laser design are:

- the three DBR mirrors maximum reflectivities (RL, RC, RR) or the corresponding lengths (Lgr,M, Lgr,C, Lgr,S); we assumed the coupling coefficient κ to be the same for all of the gratings;
- the right (LS) and left (LM) cavity lengths.

In order to obtain an estimate of the cavity fPPR, once the cavity and material parameters have been fixed, we calculate the frequency separation in the condition indicated in Figure 2b. While in this condition, the device could be self-pulsating, in proper operation conditions above-threshold, a small tuning of the cavity modes' positions will ensure an optimal intensity of the PPR peak with a small increase of fPPR, as shown in Figure 2c, producing the extra peak in the small signal modulation response due to PPR, which allows a significant extension of the laser modulation bandwidth [12,16].

In particular, the PPR frequency is computed choosing the value of the lossless left grating reflectivity RL and assuming the output right facet to be either cleaved, for realization simplicity, or realized with a grating with maximum reflectivity RR=32%. Three values of the grating coupling coefficient (κ=50, 100, 200 cm^{-1}) were considered, and the waveguides and grating losses were assumed to be 10 cm^{-1}. The remaining cavity parameters (central grating reflectivity RC and the total master and slave cavity lengths LS and LM) have been varied; for this first analysis, we assume LS=LM. Results and considerations for different values of the ratio LS/LM will be reported in the following.

The results of this PPR frequency analysis at threshold have been represented in Figure 4 in which we kept constant the left grating reflectivity RL=90%, and we considered as parameters for each figure the coupling coefficient κ. These figures are fundamental to select the structures to consider also on the basis of the available technology for the grating realization and the active material gain characteristics.

In Figure 4, the fPPR map is represented both for the case of the cleaved (continuous line) and grating (dashed line) output facet. In the figure, the dotted horizontal lines indicate the number of the cavity modes inside the central reflectivity lobe of the left grating of the master cavity r←gr,M(λ), which is the grating with the largest optical bandwidth (see Figure 1).

This parameter is important because the larger is its value, the greater is the possibility of competition between the lasing mode and the adjacent ones, which can lead to mode jumps when tuning the cavity phase. As can be seen from the maps inFigure 4, fPPR can be obtained over a broad range of frequencies, from 10 to 80GHz. As we predicted, small fPPR values are obtained for higher values of the central grating reflectivity RC due to the weak coupling between the two cavities, while the opposite is obtained for smaller values. Obviously, also the cavity length affects the value of fPPR, which decreases for longer cavities because of the reduction of the free spectral range (FSR). The regions in Figure 4's maps where the constant fPPR lines have not been reported indicate operation conditions in which, with the approximation used to generate the maps, the gain margin with respect to the other modes of the cavity is very small or lasing can be found at frequencies strongly shifted from the Bragg condition due to the complexity of the RTG reflectivity curve. The latter behavior is usually due to the higher number of cavity modes inside the main reflectivity lobe of r←. From our experience [16], in these regions, the extended modulation bandwidth conditions could still be obtained, but they are more sensitive to the cavity parameters. To simplify the reading of the maps, the truncation points have been connected with a thick dash-point blue line. As can be seen, the frequency selection introduced using a Bragg grating at the output facet instead of a cleaved surface allows one to extend the fPPR curves over all of the considered parameters' range.

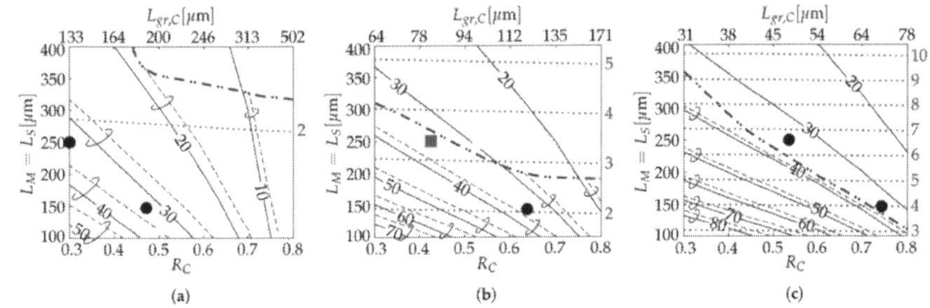

Figure 4. Coupled cavities DBR laser design maps for equal cavity length (LM=LS): constant photon-photon resonance (PPR) frequency (in GHz) curves as functions of the central grating reflectivity (RC) and for the cleaved (dashed lines map) and grating (continuous lines map) right facet with the same maximum reflectivity RR=32%. Results are obtained at the laser threshold using the transmission matrix method for a lossless left reflectivity RL=90% and with κ=50cm−1 (**a**), κ=100cm−1 (**b**) and κ= 200cm−1 (**c**). In the area above the double dotted line, where the dashed constant frequency lines are not reported, the PPR conditions are more critical to be found. The red dotted lines, referring to the right vertical axes, indicate the number of longitudinal modes of the slave cavity inside the reflectivity main lobe of the grating with the largest optical bandwidth. The large markers indicate the structures that we have analyzed in their small and large signal modulation responses; the square marker indicates the structure whose dynamic results have been reported in this paper.

The previous analysis has been repeated also for different lengths of the two sections (LS≠LM) while keeping constant the total cavity length; the results show a significant reduction of the PPR frequency and an increase of the area of mode competition as long as the master cavity length is reduced. In the following, we therefore present results only for the case of equal cavity lengths.

To quantitatively highlight the fundamental role of the central grating reflectivity on the fPPR, the map in Figure 5 shows its value when RR and RC are varied for a constant value of RL=0.90 and two values of the cavity lengths LL=LR=150 μm and 250 μm. The estimated values of fPPR show a large independence from RR and a significant dependence on κ due to the change of the effective grating length for a given RC value.

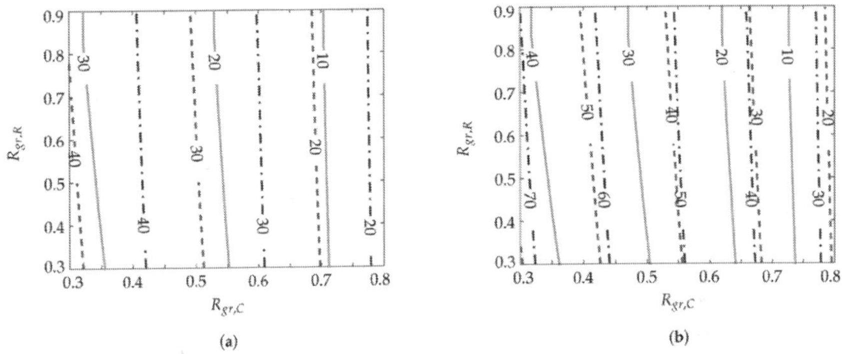

Figure 5. PPR frequency map as a function of the maximum central grating reflectivity Rgr,C and the maximum right grating reflectivity Rgr,R, for active section lengths LM=LS=150 μm (**a**) and LM=LS=250 μm (**b**). Three values of the grating coupling coefficient are considered: κ=50cm−1 (continuous lines), κ=100cm−1(dashed lines) and κ=200cm−1 (dashed-dotted lines); the labels on the lines report the PPR frequency. The left grating maximum reflectivity is 90%.

3. DYNAMIC CHARACTERISTICS

Recently, the small signal properties of the two integrated DBR laser structures have been analyzed using composite mode theory [27,28] and also experimentally realized and characterized [28,29] with respect to their small signal modulation response.

In this section, we report the results of the dynamic simulations using our FDTW numerical code [31]. These simulations have as objectives the verification of operation conditions showing the presence of the PPR in the small signal modulation response to confirm the results of the previous static design procedure in Section 2, but they also aim to show the existence of extended modulation bandwidth conditions allowing digital data transmission at a higher bit-rate with respect to one obtainable in single-cavity DBR lasers realized with the same active material.

More details about the implemented FTDW method are presented in Appendix A2.

As highlighted in Figure 2 and in Figure 3, a fine tuning of the longitudinal cavity modes' position obtained varying the phase ϕS is generally required in order to obtain the modulation bandwidth enhancement. In real devices, this tuning is generally accomplished using the phase-control sections currents $I_{M,P}$ and $I_{S,P}$ (Figure 1a). In the following simulations, for consistency with the presented analysis at threshold and in order to simplify the results interpretations, we simply represented the phase tuning adding a phase ϕS to the electric field propagating in the slave laser section (Figure 1b).

We simulated the six laser cavities that we indicated with markers in Figure 4's maps. We have chosen cavities with a PPR around 35 GHz, and between the various options, we considered the cases with short cavity lengths to limit the parasitic effects. Furthermore, we have decided to operate with cavity parameters granting only a limited number of longitudinal modes in the $r_{\leftarrow}$ main reflectivity lobe, allowing the use, for realization simplicity, of an output cleaved facet. The material and the waveguide parameters used in the simulation are reported in Table 1; the values have been chosen in agreement with [27].

For all of the considered structures, when varying the tuning phase, the obtained modulation results present a very similar behavior with respect to the injected currents in the active sections. In the map in Figure 6a, we summarize this behavior as a function of the currents I_M and I_S normalized with respect to the threshold currents $I_{th,M}$ and $I_{th,S}$ of the master and slave lasers in isolation: the former is composed of the left grating, the master active region and the central grating, while the latter is defined as the central grating, the slave active section and the right cleaved facet.

In the upper part of the map (Region (A)), in the considered structures, we obtained typically self-sustained relaxation oscillations (RO) as described, e.g., in [32] for lasers with feedback. In this region, sinusoidal self-pulsations (SP) can generally be found over wide ranges of the tuning phase when the currents in the two active sections are properly chosen: in the two examples shown in Figure 7, we report two cases of SP found when the two active section currents are the same and for two values of the phase control currents. The phase control allows one to shift the oscillation frequency from a minimum value corresponding to the PPR frequency obtained in the previously-reported design procedure to higher values. The results show in both cases the possibility to obtain a good extinction ratio and narrow RF spectra.

In Region (B), single mode operation has been typically obtained changing the control phase; examples of the corresponding small signal modulation characteristics have been reported in Figure 6b. Obviously, these operation conditions are not suitable for the extended digital modulation we are looking for.

The single mode behavior in continuous wave (CW) is also obtained in Region (D) of higher slave injection current, over a wide range of the control phase; in this region, however, the modulation results shown in Figure 6d clearly show that a large signal modulation bandwidth extension can be obtained for some values of the tuning phase ϕS.

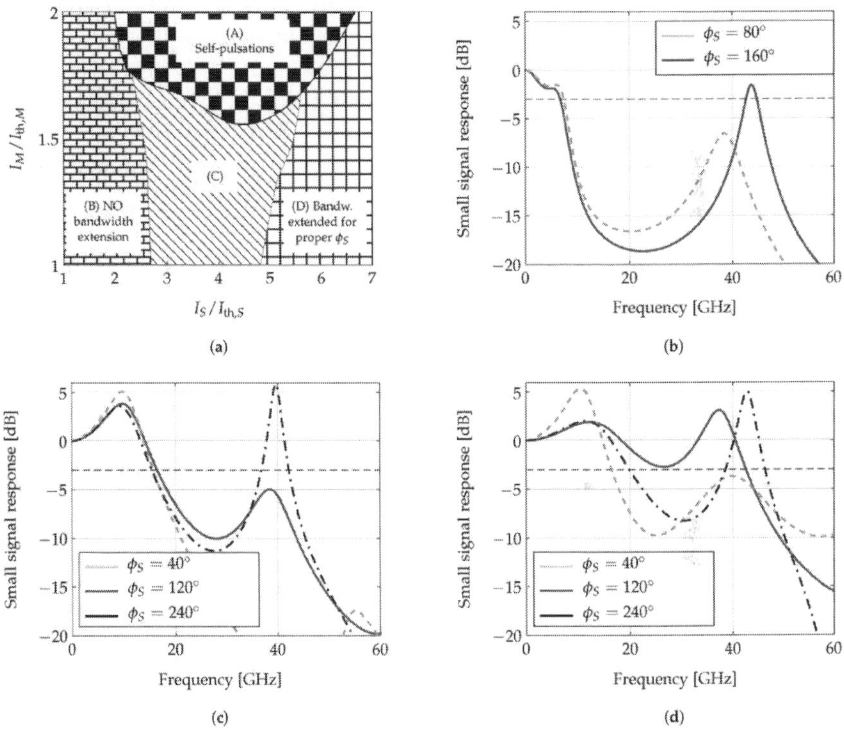

Figure 6. (**a**) Map in the normalized master and slave active sections' current plane indicating the four typical dynamic behaviors found tuning the phase control ϕS. Examples of the typical small signal modulation response obtained with different values of ϕS in the B, C, D regions are reported in (**b**,**c**,**d**) of this figure for the structure characterized by the red square marker in Figure 4b. The horizontal dashed lines represent the −3 dB level for the modulation response.

In Region (C), the operation condition in between those previously described in Regions (B) and (D) is obtained as shown in Figure 6c. The dashed modulation curve for ϕS=40∘ in Figure 6c represents an operation condition whose response is practically equal to that of a corresponding single-section DBR laser; the effect of the coupled cavity DBR configuration can be directly obtained comparing the other results with this one. Furthermore, we put in evidence that the borders between the various regions are "soft" in the sense that one should consider a gradual transition of the laser behavior from Regions (B), (C) and (D). A sharper transition has been typically found at the borders with Region (A).

We discuss now for brevity only the dynamic simulation results obtained for one of the six structures indicated with markers in Figure 4: the laser cavity indicated with a squared marker in Figure 4b, with κ=100cm−1, Lgr,L=181 μm, LM=LS=250 μm, Lgr,C=82 μm, the right facet as cleaved and having a threshold current density Jth=2.4kA/cm2. For this laser, we report in detail the results obtained only for bias points in Region (D) of Figure 6a and allowing high speed large signal modulation.

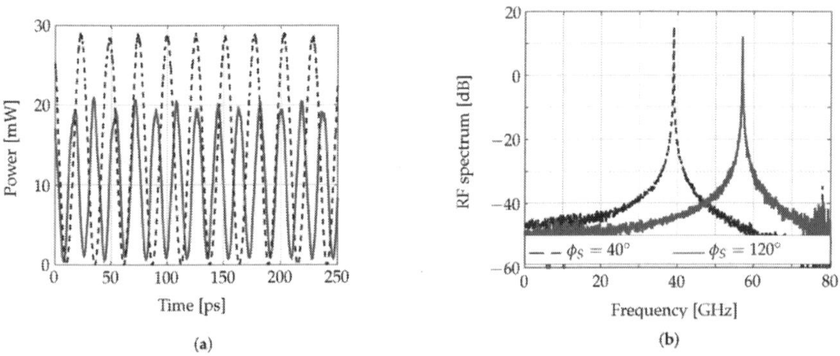

Figure 7. Two examples of sinusoidal the self-pulsation (SP) operation in Region (A) ofFigure 6a for IS=IM=30mA for two different values of ϕS: (**a**) the time domain signals and (**b**) the corresponding RF spectrum. The dashed black line (ϕs=40∘) is used for the minimum value of the PPR frequency fPPR≈39GHz, while the continuous red line (ϕs=120∘) is used for the case fPPR≈57GHz.

The small signal modulation results are show in Figure 8's color map, where the modulation frequency is reported in the abscissa, while the ordinates indicate the static phase change ϕS in the slave section used to tune the position of the cavity modes to obtain an operation condition allowing the extension of the modulation bandwidth.

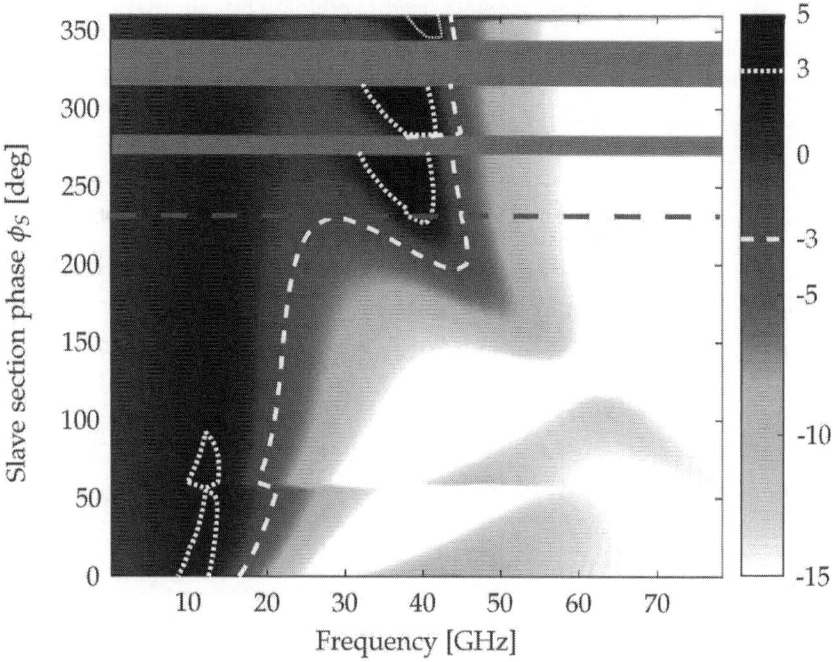

Figure 8. Modulation response map, in dB, for the device indicated by the red square marker in Figure 4b, with, in the ordinate, the tuning phase shift ϕS introduced in the slave optical cavity. The output power is approximately 25 mW; IS=60mA, IM=8mA. Contour lines indicate -3 dB (dashed line) and +3 dB (dotted line) levels of the modulation response. The red dashed horizontal line indicates the operation condition chosen for the large signal simulation.

The two gray horizontal strips in the map represent regions with self-sustained RO similar to those in Region (A) shown in Figure 6, but with a smaller extinction ratio corresponding to the presence of a mode in the optical spectrum with a side mode suppression ratio greater than 20 dB. In these operation conditions, the large signal modulation eye diagram is practically close also for a low bit rate. The blue "horn" pointing to the lower part of the map shows the effect of the presence of PPR on the modulation response; for high frequency PPR, the resonance peak is weak and becomes stronger as the mode separation decreases. In the lower part of the map, this effect does not appear, because in this range of phases ϕS, the closest side mode is on the shorter wavelength side with respect to the lasing mode: in this condition, the PPR effect does not take place [12].

Table 1. Main material parameters used for the FDTW simulations.

Symbol	Description	Value
$n_{eff,0}$	Effective refractive index	3.2
v_g	Group velocity	8.3×10^7 m/s
Γ_{xy}	Transversal optical confinement factor	10%
N_0	Carrier density at transparency	1×10^{18} cm^{-3}
a_0	Active region differential gain	13×10^{-16} cm^2
α_i	Material intrinsic losses	10 cm^{-1}
λ_B	Bragg wavelength	1.550 µm
α_{LEF}	Linewidth enhancement factor	3
ε	Non-linear gain compression factor	5×10^{-17} cm^3
A	Mono-molecular recombination coefficient	1.4 ns^{-1}
B	Bi-molecular recombination coefficient	0.1 cm^3 ns^{-1}
C	Auger recombination coefficient	7.5 cm^6 ns^{-1}

The horizontal discontinuities appearing in the lower part of the modulation response when varying the tuning phase, as for example at ϕS=65∘ in Figure 8, are associated with longitudinal mode jumps. This behavior highlights the importance of limiting the number of longitudinal modes inside the main lobe of the reflectivity peak.

For the value of the slave cavity phase highlighted by the red dashed horizontal line, where the modulation bandwidth is extended with not a too high resonance peak, the value of the PPR frequency is around 40 GHz, in good agreement with thefPPR value used to choose the cavity parameters from the map in Figure 4; this result, valid for all of the simulated devices, confirms the reliability of the proposed design procedure based on the PPR analysis at threshold.

The laser operation condition in Figure 8 was considered for the large signal modulation analysis using an FDTW approach. For this analysis, the master section current was kept constant, and a NRZ pseudo-random bit sequence (PRBS) composed of 215−1 bits was applied to the slave section, and the slave section phase ϕS was set to 240∘ in order to take advantage of the bandwidth extension provided by the PPR effect.

Results are presented as eye diagrams in Figure 9. For each eye, on the right axes, we report the output power *P*, while on the left vertical axes, we indicate *P* normalized with respect to the bit "0" and "1" levels P0 and P1: p=(P−P0)/(P1−P0). In order to allow for an easy estimation of the eyes opening, in the figure, we also report the limits for optical transmission systems indicated by the IEEE P802.3ba standard [33].

As expected, the results reported in Figure 9a,b show that open eyes can be obtained using a 40 Gbit/s bit-rate when P1/P0=6 dB (a) and P1/P0=3 dB (b). Similar results were obtained in a wide range from 150° to 270° of the tuning phase shift ϕS. For a large-scale deployment of these devices, a control electronic circuit with a lookup table could be used for the correct selection of the currents in the phase control regions to ensure this extended modulation bandwidth operation.

We then increased the bit-rate and found that, with respect to the considered mask, the upper limit allowed one to obtain an open eye diagram as 60 Gbit/s when operating with P1/P0=6 dB (c) and 80 Gbit/s when P1/P0=3 dB (d), which indicates an extension of the modulation frequency well above the CPR frequency. These results were obtained in a significantly reduced range of the tuning phase shift ϕS extending from 240° to 260°. The eyes are closed for any value of ϕS when higher bit rates are used.

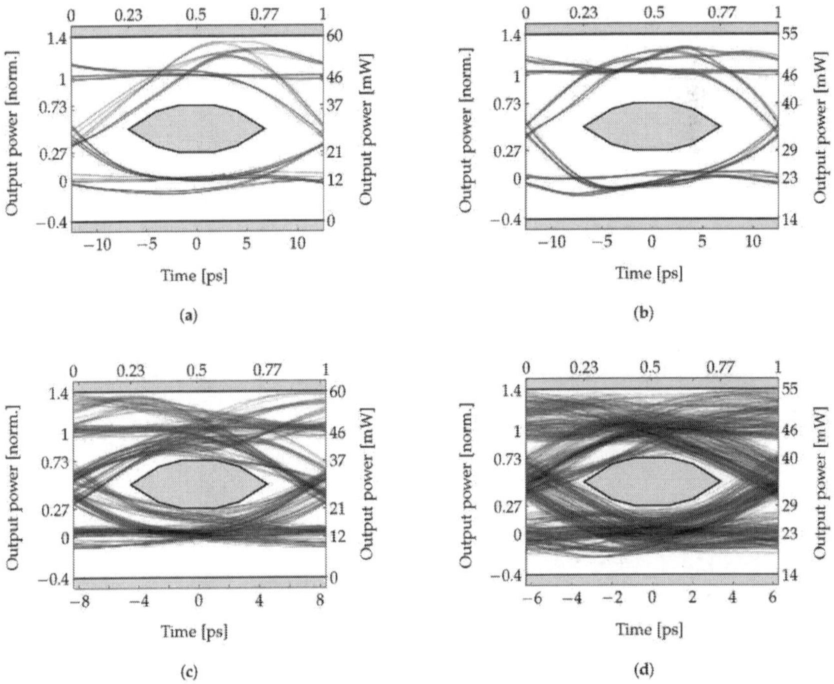

Figure 9. Eye diagrams for the operation condition considered in Figure 8 using ϕS=240°, computed with a 215−1 bits-long non-return-to-zero

(NRZ) pseudo-random bit sequence (PRBS) sequence. The master section current is fixed (IM=8 mA), while the slave section current is modulated from IS,0 (corresponding to the logic "0" level) to IS,1 (corresponding to the logic "1" level). (**a**) IS,0=40 mA, IS,1=110 mA, corresponding to a 6 modulation amplitude, with a 40 Gbit/s bitrate, increases to 60 Gbit/s in (**c**). (**b**) IS,0=60 mA, IS,1=110 mA, corresponding to a 3 dB modulation amplitude, with a 40 Gbit/s bitrate, increased to 80 Gbit/s in (**d**).

4. CONCLUSIONS

We proposed a design procedure to exploit the PPR effect in an integrated device composed of two mutually-coupled DBR lasers. The design maps, obtained using the transmission matrix method, allow for an optimal combination of the cavity parameters once the required bandwidth extension and the technological aspects have been defined.

A detailed investigation of the dynamics of this structure using an FDTW simulator over a wide range of the bias conditions allows one to obtain a tunable self-pulsating signal with a good extinction ratio and extended modulation conditions in the range of the proposed design procedure. The approach gives the cavity modes' position, resulting in an increase of the −3 dB modulation bandwidth thanks to the PPR peak appearing close to the design frequency. Finally, the large signal modulations with PRBS signals allowed one to obtain eye diagrams, which are open up to 60Gbits−1.

ACKNOWLEDGMENTS

This work is partially supported by the Sandia LDRD program, funded by the U.S. Department of Energy under Contract No. DE-AC04-94AL85000 and by the EU FP7 project "Development of low-cost technologies for the fabrication of high-performance telecommunication lasers" (DeLight) under Grant 224366.

AUTHOR CONTRIBUTIONS

W.W.C. proposed the mutually injected DBR lasers device investigated in the paper; P.B. implemented the numerical methods and the analysis tools, and performed the simulations; P.B. and I.M. analyzed the simulation results; P.B., W.W.C. and I.M. wrote the paper.

CONFLICTS OF INTEREST

The authors declare no conflict of interest.

REFERENCES

1 De Valicourt, G.; Levaufre, G.; Pointurier, Y.; Le Liepvre, A.; Antona, J.C.; Jany, C.; Accard, A.; Lelarge, F.; Make, D.; Duan, G.H. Direct Modulation of Hybrid Integrated InP/Si Transmitters for short and long reach access network. *J. Lightw. Technol.* **2015**, *33*, 1608–1616.

2 Coldren, L.A.; Corzine, S.W.; Maǎnović, M.L. *Diode Lasers and Photonic Integrated Circuits*; Wiley: New York, NY, USA, 2012.

3 Montrosset, I.; Bardella, P. Laser dynamics providing enhanced-modulation bandwidth. *Proc. SPIE* **2014**, *9134*.

4 Vahala, K.; Yariv, A. Detuned loading in coupled cavity semiconductor lasers—effect on quantum noise and dynamics. *Appl. Phys. Lett.* **1984**, *45*, 501–503.

5 Vahala, K.; Paslaski, J.; Yariv, A. Observation of modulation speed enhancement, frequency modulation suppression, and phase noise reduction by detuned loading in a coupled-cavity semiconductor laser. *Appl. Phys. Lett.* **1985**, *46*, 1025–1027.

6 Feiste, U. Optimization of modulation bandwidth in DBR lasers with detuned Bragg reflectors. *IEEE J. Quant. Electron.* **1998**, *34*, 2371–2379.

7 Kjebon, O.; Schatz, R.; Lourdudoss, S.; Nilsson, S.; StAlnacke, B.; Backbom, L. Two–section InGaAsP DBR-lasers at 1.55 μm wavelength with 31 GHz direct modulation bandwidth. In Proceedings of the International Conference on Indium Phosphide and Related Materials, Hyannis, MA, USA, 11–15 May 1997; pp. 665–668.

8 Chaciński, M.; Schatz, R. Impact of Losses in the Bragg Section on the Dynamics of Detuned Loaded DBR Lasers. *IEEE J. Quant. Electron.* **2010**, *46*, 1360–1367.

9 Pan, X.; Olesen, H.; Tromborg, B. Modulation characteristics of tunable DFB/DBR lasers with one or two passive tuning sections. *IEEE J. Quant. Electron.* **1989**, *25*, 1254–1260.

10 Kjebon, O.; Schatz, R.; Lourdudoss, S.; Nilsson, S.; StAlnacke, B.; Backbom, L. 30 GHz direct modulation bandwidth in detuned loaded InGaAsP DBR lasers at 1.55 μm wavelength. *Electron. Lett.* **1997**, *33*, 488–489.

11 Bach, L.; Kaiser, W.; Reithmaier, J.P.; Forchel, A.; Berg, T.; Tromborg, B. Enhanced direct–modulated bandwidth of 37 GHz by a multi–section laser with a coupled–cavity–injection–grating design. *Electron. Lett.* **2003**, *39*, 1592–1593.

12 Bardella, P.; Montrosset, I. A New Design Procedure for DBR Lasers Exploiting the Photon–Photon Resonance to Achieve Extended Modulation Bandwidth. *IEEE J. Sel. Top. Quant. Electron.* **2013**, *19*.

13 Reithmaier, J.P.; Kaiser, W.; Bach, L.; Forchel, A.; Feies, V.; Gioannini, M.; Montrosset, I.; Berg, T.; Tromborg, B. Modulation speed enhancement by coupling to higher order resonances: a road towards 40 GHz bandwidth lasers on InP. In Proceedings of 2005 International Conference on Indium Phosphide and Related Materials, Glasgow, UK, 8–12 May 2005; pp. 118–123.

14 Reithmaier, J.; Bach, L.; Kaiser, W. Multisectional Laser. US Patent 7,570,681, 2009.

15 Gerschütz, F.; Fischer, M.; Koeth, J.; Krestnikov, I.; Kovsh, A.; Schilling, C.; Kaiser, W.; Höfling, S.; Forchel, A. 1.3 μm Quantum Dot Laser in coupled-cavity-injection-grating design with bandwidth of 20 GHz under direct modulation.*Opt. Express* **2008**, *16*, 5596–5601.

16 Vallone, M.; Bardella, P.; Montrosset, I. Enhanced Modulation Bandwidth in Complex Cavity Injection Grating Lasers.*IEEE J. Quant. Electron.* **2011**, *47*, 1269–1276.

17 Radziunas, M.; Glitzky, A.; Bandelow, U.; Wolfrum, M.; Troppenz, U.; Kreissl, J.; Rehbein, W. Improving the Modulation Bandwidth in Semiconductor Lasers by Passive Feedback. *IEEE J. Sel. Top. Quant. Electron.* **2007**, *13*, 136–142.

18 Morthier, G.; Schatz, R.; Kjebon, O. Extended modulation bandwidth of DBR and external cavity lasers by utilizing a cavity resonance for equalization. *IEEE J. Quant. Electron.* **2000**, *36*, 1468–1475

19 Grillot, F.; Wang, C.; Naderi, N.; Even, J. Modulation Properties of Self-Injected Quantum-Dot Semiconductor Diode Lasers. *IEEE J. Sel. Top. Quant. Electron.* **2013**, *19*.

20 Simpson, T.B.; Liu, J.; Gavrielides, A. Bandwidth enhancement and broadband noise reduction in injection-locked semiconductor lasers. *IEEE Photonics Technol. Lett.* **1995**, *7*, 709–711.

21 Murakami, A.; Kawashima, K.; Atsuki, K. Cavity resonance shift and bandwidth enhancement in semiconductor lasers with strong light injection. *IEEE J. Quant. Electron.* **2003**, *39*, 1196–1204.

22 Lau, E.; Wong, L.J.; Wu, M. Enhanced Modulation Characteristics of Optical Injection-Locked Lasers: A Tutorial. *IEEE J. Sel. Top. Quant. Electron.* **2009**, *15*, 618–633.

23 Enard, A.; Resneau, P.; Calligaro, M.; Parillaud, O.; Krakowski, M.; Vallone, M.; Bardella, P.; Montrosset, I. Mode locking and bandwidth enhancement in single section ridge laser with two spatial modes. In Proceedings of the 22nd IEEE International Semiconductor Laser Conference (ISLC), Kyoto, Japan, 26–30 September 2010; pp. 113–114.

24 Dalir, H.; Koyama, F. High-speed operation of bow-tie-shaped oxide aperture VCSELs with photon–photon resonance. *Appl. Phys. Express* **2014**, *7*.

25 Dalir, H.; Koyama, F. Highly stable operations of transverse-coupled cavity VCSELs with enhanced modulation bandwidth. *Electron. Lett.* **2014**, *50*, 823–824.

26 Gerschütz, F.; Fischer, M.; Koeth, J.; Krestnikov, I.; Kovsh, A.; Schilling, C.; Kaiser, W.; Höfling, S.; Forchel, A. 1.3 μm Quantum Dot Laser in coupled-cavity-injection-grating design with bandwidth of 20 GHz under direct modulation.*Opt. Express* **2008**, *16*, 5596–5601.

27 Chow, W.; Yang, Z.; Vawter, G.; Skogen, E. Modulation response improvement with isolator-free injection-locking.*IEEE Photonics Technol. Lett.* **2009**, *21*, 839–841.

28 Yang, Z.; Tauke-Pedretti, A.; Vawter, G.; Chow, W. Mechanism for Modulation Response Improvement in Mutually Injection-Locked Semiconductor Lasers. *IEEE J. Quant. Electron.* **2011**, *47*, 300–305.

29 Tauke-Pedretti, A.; Vawter, G.; Skogen, E.; Peake, G.; Overberg, M.; Alford, C.; Chow, W.; Yang, Z.; Torres, D.; Cajas, F. Mutual Injection Locking of Monolithically Integrated Coupled-Cavity DBR Lasers. *IEEE Photonics Technol. Lett.***2011**, *23*, 908–910.

30 Tauke-Pedretti, A.; Skogen, E.; Vawter, G.; Chow, W. Mutually Injection Locked Lasers for Enhanced Frequency Eesponse. US Patent 8,687,665, 2014.

31 Bardella, P.; Montrosset, I. Analysis of self-pulsating three–section DBR lasers. *IEEE J. Sel. Top. Quant. Electron.* **2005**,*11*, 361–366.

32 Mork, J.; Tromborg, B.; Mark, J. Chaos in semiconductor lasers with optical feedback: Theory and experiment. *IEEE J. Quant. Electron.* **1992**, *28*, 93–108.

33 802.3-2008 IEEE Standard for Information Technology-Specific Requirements-Part 3: Carrier Sense Multiple Access with Collision Detection (CMSA/CD) Access Method and Physical Layer Specifications. Available online: http://ieeexplore.ieee.org/ servlet/ opac?punumber=4726157 (accessed on 29 December 2015).

34 Bonello, R.; Montrosset, I. Analysis of multisection and multielectrode semiconductor lasers. *J. Lightw. Technol.* **1992**,*10*, 1890–1900.

35 Kim, B.S.; Chung, Y.; Lee, J.S. An efficient split-step time-domain dynamic modeling of DFB/DBR laser diodes. *IEEE J. Quantum Electron.* **2000**, *36*, 787–794.

36 Jones, D.J.; Zhang, L.; Carroll, J.; Marcenac, D.D. Dynamics of monolithic passively mode-locked semiconductor lasers. *IEEE J. Quantum Electron.* **1995**, *31*, 1051–1058.

Index